TREES

OF ALABAMA

Philip Henry Gosse as a young man of twenty-nine, the year of his return to England from Alabama, painted by his brother, William Gosse. (1839, watercolor on ivory, courtesy of the National Portrait Gallery—London)

Philip Henry Gosse (1810–1888) was an English naturalist and illustrator who spent eight months of 1838 on the Alabama frontier, teaching planters' children in Dallas County and studying the native flora and fauna. Years after returning to England, he published the now-classic *Letters from Alabama: Chiefly Relating to Natural History*, with twenty-nine important black-and-white illustrations included. He also produced, during his Alabama sojourn, forty-nine remarkable watercolor plates of various plant and animal species, mainly insects, now available in *Philip Henry Gosse: Science and Art in "Letters from Alabama" and "Entomologia Alabamensis."*

The Gosse Nature Guides are a series of natural history guidebooks prepared by experts on the plants and animals of Alabama and designed for the outdoor enthusiast and ecology layman. Because Alabama is one of the nation's most biodiverse states, its residents and visitors require accurate, accessible field guides to interpret the wealth of life that thrives within the state's borders. The Gosse Nature Guides are named to honor Philip Henry Gosse's early appreciation of Alabama's natural wealth and to highlight the valuable legacy of his recorded observations. Look for other volumes in the Gosse Nature Guides series at http://uapress.ua.edu.

TREES

OF ALABAMA

LISA J. SAMUELSON

with Photographs by Michael E. Hogan

THE UNIVERSITY OF ALABAMA PRESS   TUSCALOOSA

The University of Alabama Press
Tuscaloosa, Alabama 35487-0380
uapress.ua.edu

Typeface: Scala Pro and Scala Sans Pro

Cover images: Clockwise from top left corner, Florida maple leaves,
scarlet oak leaves, witch-hazel leaves, sugar maple fruit, and pond
pine serotinous seed cones; courtesy of Michael E. Hogan
Cover and text design: Michele Myatt Quinn

Photographs and drawings by Michael E. Hogan, except where noted.

Portions of this book were previously published by Pearson Education,
Inc. as *Forest Trees: A Guide to the Southeastern and Mid-Atlantic
Regions of the United States* by Lisa J. Samuelson and Michael E. Hogan.

Publication made possible in part by a generous contribution from
the Alabama Museum of Natural History, Tuscaloosa, Alabama.

Library of Congress Cataloging-in-Publication Data
Names: Samuelson, Lisa J., author. | Hogan, Michael E., photographer.
Title: Trees of Alabama / Lisa J. Samuelson; photographs by Michael E. Hogan.
Other titles: Gosse nature guides.
Description: Tuscaloosa : The University of Alabama Press, [2019] | Series:
    Gosse nature guides | Includes bibliographical references and index.
Identifiers: LCCN 2019003188| ISBN 9780817359416 (pbk.)
    | ISBN 9780817392307
Subjects: LCSH: Tree —Alabama—Identification.
Classification: LCC QK145 .S26 2019 | DDC 582.1609761—dc23
    LC record available at https://lccn.loc.gov/2019003188

We emerge into the high road, bounded on both sides by the hard-wood forest, where the oaks and hickories, the sycamore and the tulip-tree, the chestnut and the sweet-gum, cast a greenwood shade, varied, however, now with gorgeous tints, like the rays that stream through the painted window of some old cathedral, by the dying foliage.

—Philip Henry Gosse, *Letters from Alabama*, Letter XVI,
   November 1st, 1838

Contents

Introduction   1

1 How to Identify Trees   3

2 Identification Features   7

3 Guide to the Identification of Trees   9

Guide to Gymnosperms   9

Guides to Angiosperms   11

4 Species Accounts   31

Cypress Family (*Cupressaceae*)

Atlantic White-Cedar (*Chamaecyparis thyoides*)   31

Eastern Redcedar (*Juniperus virginiana* var. *virginiana*)   33

Pondcypress (*Taxodium ascendens*)   35

Baldcypress (*Taxodium distichum*)   37

Pine Family (*Pinaceae*)

Shortleaf Pine (*Pinus echinata*)   39

Slash Pine (*Pinus elliottii*)   41

Spruce Pine (*Pinus glabra*)   43

Longleaf Pine (*Pinus palustris*)   45

Pond Pine (*Pinus serotina*)   47

Eastern White Pine (*Pinus strobus*)   49

Loblolly Pine (*Pinus taeda*)   51

Virginia Pine (*Pinus virginiana*)   53

Eastern Hemlock (*Tsuga canadensis*)   55

Moscatel Family (*Adoxaceae*)

Rusty Blackhaw (*Viburnum rufidulum*)   57

Sweetgum Family (*Altingiaceae*)

    Sweetgum (*Liquidambar styraciflua*)   59

Cashew Family (*Anacardiaceae*)

    American Smoketree (*Cotinus obovatus*)   61

    Winged Sumac (*Rhus copallinum*)   63

    Smooth Sumac (*Rhus glabra*)   65

    Poison-Sumac (*Toxicodendron vernix*)   67

Custard-Apple Family (*Annonaceae*)

    Pawpaw (*Asimina triloba*)   69

Holly Family (*Aquifoliaceae*)

    Large Gallberry (*Ilex coriacea*)   71

    Possumhaw (*Ilex decidua*)   73

    American Holly (*Ilex opaca*)   75

    Yaupon (*Ilex vomitoria*)   77

Ginseng Family (*Araliaceae*)

    Devil's Walkingstick (*Aralia spinosa*)   79

Birch Family (*Betulaceae*)

    Hazel Alder (*Alnus serrulata*)   81

    Black Birch (*Betula lenta*)   83

    River Birch (*Betula nigra*)   85

    Hornbeam (*Carpinus caroliniana*)   87

    Hophornbeam (*Ostrya virginiana*)   89

Catalpa Family (*Bignoniaceae*)

    Southern Catalpa (*Catalpa bignonioides*)   91

Strawberry-Shrub Family (*Calycanthaceae*)

    Sweetshrub (*Calycanthus floridus*)   93

Cannabis and Hop Family (*Cannabaceae*)

    Sugarberry (*Celtis laevigata*)   95

Hackberry (*Celtis occidentalis*)   97

Georgia Hackberry (*Celtis tenuifolia*)   99

Dogwood Family (*Cornaceae*)

Alternate-Leaf Dogwood (*Cornus alternifolia*)   101

Flowering Dogwood (*Cornus florida*)   103

Water Tupelo (*Nyssa aquatica*)   105

Swamp Tupelo (*Nyssa biflora*)   107

Blackgum (*Nyssa sylvatica*)   109

Ebony Family (*Ebenaceae*)

Common Persimmon (*Diospyros virginiana*)   111

Heath Family (*Ericaceae*)

Sourwood (*Oxydendrum arboreum*)   113

Sparkleberry (*Vaccinium arboreum*)   115

Spurge Family (*Euphorbiaceae*)

Chinese Tallowtree (*Triadica sebifera*)   117

Bean or Pea Family (*Fabaceae*)

Mimosa (*Albizia julibrissin*)   119

Eastern Redbud (*Cercis canadensis*)   121

Honeylocust (*Gleditsia triacanthos*)   123

Black Locust (*Robinia pseudoacacia*)   125

Beech Family (*Fagaceae*)

American Chestnut (*Castanea dentata*)   127

Allegheny Chinkapin (*Castanea pumila*)   129

American Beech (*Fagus grandifolia*)   131

White Oak (*Quercus alba*)   133

Bluff Oak (*Quercus austrina*)   135

Scarlet Oak (*Quercus coccinea*)   137

Durand Oak (*Quercus durandii*)   139

Southern Red Oak (*Quercus falcata*)   141

Laurel Oak (*Quercus hemisphaerica*)   143

Bluejack Oak (*Quercus incana*)   145

Turkey Oak (*Quercus laevis*)   147

Swamp Laurel Oak (*Quercus laurifolia*)   149

Overcup Oak (*Quercus lyrata*)   151

Sand Post Oak (*Quercus margarettiae*)   153

Blackjack Oak (*Quercus marilandica*)   155

Swamp Chestnut Oak (*Quercus michauxii*)   157

Chestnut Oak (*Quercus montana*)   159

Chinkapin Oak (*Quercus muehlenbergii*)   161

Water Oak (*Quercus nigra*)   163

Cherrybark Oak (*Quercus pagoda*)   165

Willow Oak (*Quercus phellos*)   167

Northern Red Oak (*Quercus rubra*)   169

Shumard Oak (*Quercus shumardii*)   171

Post Oak (*Quercus stellata*)   173

Nuttall Oak (*Quercus texana*)   175

Black Oak (*Quercus velutina*)   177

Live Oak (*Quercus virginiana*)   179

Walnut Family (*Juglandaceae*)

Water Hickory (*Carya aquatica*)   181

Bitternut Hickory (*Carya cordiformis*)   183

Pignut Hickory (*Carya glabra*)   185

Pecan (*Carya illinoinensis*)   187

Shellbark Hickory (*Carya laciniosa*)   189

Nutmeg Hickory (*Carya myristiciformis*)   191

Red Hickory (*Carya ovalis*)   193

Shagbark Hickory (*Carya ovata*)   195

Sand Hickory (*Carya pallida*)   197

Mockernut Hickory (*Carya tomentosa*)   199

Butternut (*Juglans cinerea*)   201

Black Walnut (*Juglans nigra*)   203

Laurel Family (*Lauraceae*)

Redbay (*Persea borbonia*)   205

Sassafras (*Sassafras albidum*)   207

Magnolia Family (*Magnoliaceae*)

Tulip-Poplar (*Liriodendron tulipifera*)   209

Cucumbertree (*Magnolia acuminata*)   211

Southern Magnolia (*Magnolia grandiflora*)   213

Bigleaf Magnolia (*Magnolia macrophylla*)   215

Umbrella Magnolia (*Magnolia tripetala*)   217

Sweetbay Magnolia (*Magnolia virginiana*)   219

Hibiscus or Mallow Family (*Malvaceae*)

Basswood (*Tilia americana*)   221

Mahogany Family (*Meliaceae*)

Chinaberry (*Melia azedarach*)   223

Fig Family (*Moraceae*)

Osage-Orange (*Maclura pomifera*)   225

Red Mulberry (*Morus rubra*)   227

Wax-Myrtle Family (*Myricaceae*)

Southern Bayberry (*Morella cerifera*)   229

Olive Family (*Oleaceae*)

Fringe-Tree (*Chionanthus virginicus*)   231

White Ash (*Fraxinus americana*)   233

Green Ash (*Fraxinus pennsylvanica*)   235

Devilwood (*Osmanthus americanus*)   237

Sycamore Family (*Platanaceae*)

  Sycamore (*Platanus occidentalis*)   239

Buckthorn Family (*Rhamnaceae*)

  Carolina Buckthorn (*Frangula caroliniana*)   241

Rose Family (*Rosaceae*)

  Downy Serviceberry (*Amelanchier arborea*)   243

  Hawthorns (*Crataegus* spp.)   245

  Southern Crabapple (*Malus angustifolia*)   247

  Chickasaw Plum (*Prunus angustifolia*)   249

  Cherry Laurel (*Prunus caroliniana*)   251

  Mexican Plum (*Prunus mexicana*)   253

  Black Cherry (*Prunus serotina*)   255

Rue Family (*Rutaceae*)

  Hercules'-Club (*Zanthoxylum clava-herculis*)   257

Willow or Poplar Family (*Salicaceae*)

  Eastern Cottonwood (*Populus deltoides*)   259

  Black Willow (*Salix nigra*)   261

Soapberry Family (*Sapindaceae*)

  Yellow Buckeye (*Aesculus flava*)   263

  Ohio Buckeye (*Aesculus glabra*)   265

  Florida Maple (*Acer floridanum*)   267

  Chalk Maple (*Acer leucoderme*)   269

  Boxelder (*Acer negundo*)   271

  Red Maple (*Acer rubrum* var. *rubrum*)   273

  Silver Maple (*Acer saccharinum*)   275

  Sugar Maple (*Acer saccharum*)   277

Sapodilla Family (*Sapotaceae*)

  Gum Bumelia (*Sideroxylon lanuginosum*)   279

  Buckthorn Bumelia (*Sideroxylon lycioides*)   281

Star-Vine Family (*Schisandraceae*)

    Anise-Tree (*Illicium floridanum*)   283

Figwort Family (*Scrophulariaceae*)

    Royal Paulownia (*Paulownia tomentosa*)   285

Quassia Family (*Simaroubaceae*)

    Tree-of-Heaven (*Ailanthus altissima*)   287

Bladdernut Family (*Staphyleaceae*)

    American Bladdernut (*Staphylea trifolia*)   289

Storax Family (*Styracaceae*)

    Carolina Silverbell (*Halesia tetraptera*)   291

    American Snowbell (*Styrax americanus*)   293

    Bigleaf Snowbell (*Styrax grandifolius*)   295

Sweetleaf Family (*Symplocaceae*)

    Sweetleaf (*Symplocos tinctoria*)   297

Tea Family (*Theaceae*)

    Loblolly-Bay (*Gordonia lasianthus*)   299

Elm Family (*Ulmaceae*)

    Water-Elm (*Planera aquatica*)   301

    Winged Elm (*Ulmus alata*)   303

    American Elm (*Ulmus americana*)   305

    Slippery Elm (*Ulmus rubra*)   307

Appendix: Winter Twigs   309

Glossary   323

Bibliography   329

Index   333

TREES

OF ALABAMA

Alabama Forest Types

Introduction

Alabama forests are part of the deciduous forest biome of the eastern United States. Three general forest associations are found in Alabama: 1) the Mixed Mesophytic Association, which consists of diverse forests in the southern Appalachian Mountains; 2) the Oak-Hickory Association, which covers most of the state and is dominated by pine or mixed-pine-hardwood forests; and 3) the Southern Mixed Hardwoods Association in the Southern Coastal Plain, which consists of many diverse types of vegetation (Vankat 1979). The United States Department of Agriculture (USDA) Forest Service Forest Inventory and Analysis National Program identified 46 forest types in Alabama, which covered 22.8 million acres of the state in 2013 (Alabama Forestry Commission 2010 and 2013). Loblolly pine is the most dominant forest type in Alabama, making up 33.8% of the forested land area, followed by mixed upland hardwood forests (12.1%), mixed loblolly pine-hardwood forests (9.6%), white oak-red oak-hickory forests (7.4%), sweetgum-tulip-poplar forests (4.8%), sweetgum-Nutall oak-willow oak forests (4.3%), sweetbay-swamp tupelo-red maple forests (3.0%), longleaf pine forests (3.0%), and slash pine forests (2.0%) (Alabama Forestry Commission 2010). The remaining 37 forest types each make up less than 2% of the total forested area. Collectively, 57% of the forest land in the state is hardwood forest or mixed-pine hardwood forest. Approximately 200 tree species are found in Alabama. This book covers most of those trees and some woody shrubs that may attain tree size.

Common names are often derived from habitat, distinctive features, locality, and use, and can also be in commemoration or an adaptation from another language (Harlow et al. 1996). Common names may be hyphenated or one word when a tree is not a true member of the family or genus (e.g., Osage-orange is not an orange, boxelder is not related to the elders). Because there can be so many common names for one species, a scientific naming system was developed to minimize confusion and group similar species together. Scientific names may be commemorative, descriptive, fanciful, or based on uses, locality, or the original name. The scientific names in this book were based on several sources (Angiosperm Phylogeny Group III 2009; Weakley 2012; Keener et al. 2018). However, scientific names and families do change over time and may vary depending on acceptance by the author. I have indicated in the text where this may be the case. *Silvics of North America*, volumes 1 and 2, edited by Burns and Honkala (1990), were important references for describing forest associates and tree sizes.

The majority of distribution maps were redrawn from the USDA Natural Resources Conservation Service PLANTS Database (USDA 2013), in which county data are based primarily on the literature, herbarium specimens, and confirmed observations. However, not all populations have been documented so many gaps in the distributions are not real. Therefore, the distribution maps should be used as general guides. Distribution maps for Durand oak, sand post oak, pignut hickory, red hickory, hackberry, Georgia hackberry, laurel oak, swamp laurel oak, and Carolina silverbell were redrawn from the Alabama Plant Atlas (Keener et al. 2018), which is based on vouchered plant specimens only.

1

How to Identify Trees

The guides included in this book are intended to help identify families and species within families. To use the guides, select the options from those provided that most closely matches the species you are observing. The choices are based on the most prominent plant characteristics. Note that bark characteristics included in the guides describe mature tree bark.

Gymnosperms

Gymnosperms include the pines, firs, spruces, larches, cypresses, hemlocks, yews, junipers, and cedars. The first step in identification of gymnosperms using leaves is to determine if leaves are needlelike (pines, firs, spruces, larches, baldcypress, hemlocks, yews) or scalelike (cedars, junipers, pondcypress). If needlelike, determine if needles are in bundles (fascicled, the pines) or unbundled (in Alabama, baldcypress and eastern hemlock). The number of needles in a fascicle, their length, and whether they are twisted, stiff, or flexible will aid in identification.

To identify a tree based on the cone, determine if the cone scales are flat (in Alabama, the pines and eastern hemlock) or peltate (the cypresses and Atlantic white-cedar). If the cone is fleshy, the species is eastern redcedar. To separate the cones of the pine species, examine the size and color of the cone and sharpness of the prickle on the cone scale.

Angiosperms

When first learning to identify Angiosperm trees in the summer, it is best to use a series of identification features beginning with the most general and ending with the most specific features. First, determine whether the specimen has an opposite, whorled, or an alternate leaf arrangement by examining how the petioles are arranged on the twig. Most species are alternate, so if a tree is opposite or whorled, that

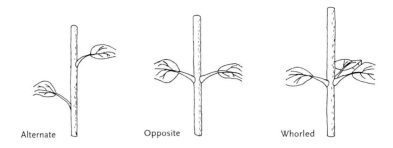

Alternate Opposite Whorled

narrows your options. Some of the most common opposite species are maples, ashes, dogwoods, viburnums, and buckeyes. Southern catalpa is a common whorled species. Next, determine if the tree has simple or compound leaves. Most species are simple, so a compound arrangement also narrows your options. An opposite and compound species is least common and includes the ashes, boxelder, buckeyes, and bladdernut. Leaf characteristics (lobing, size, shape, margin, apex, base, presence of hair or glands, venation, and odor when crushed) can then be used. In the text, leaf sizes include the blade plus the petiole unless noted otherwise. When using bark to identify trees, remember that mature tree bark is different from juvenile tree bark. If flowers or fruit are present, they can aid in identification.

In winter, bark and twig features and fruit, if fruit persists, are used in identification. A hand lens is very useful in seeing bud and leaf scar features. First, examine the arrangement of small branches

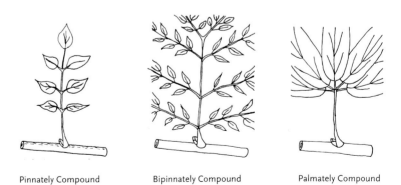

Pinnately Compound Bipinnately Compound Palmately Compound

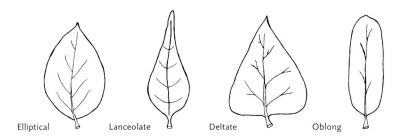

Elliptical   Lanceolate   Deltate   Oblong

or leaf scars to determine if your tree is opposite or alternate. It's best to examine the portion of a young twig that grew the past summer to see the best features. The leaf scar is where the petiole was once attached. The dots within the leaf scar are the bundle scars, which were made by the xylem that transported water in and phloem that transported sugars out of the leaf. The shape of the leaf scar and number and arrangement of bundle scars are unique. Examine the terminal bud and determine whether there is more than one, such as in the oaks, or just one and whether the single bud is positioned at an angle to the twig end (a pseudoterminal bud). Explore whether buds are round or pointed, small or large, and smooth, hairy, or glandular. Determine whether bud scales overlap (imbricate) or do not overlap (valvate). If scales are not visible, usually because the bud is hairy, then the bud is described as naked. Examine the texture and color of the hair on buds. Determine if the twig has hair, glands, thorns, or an odor when crushed.

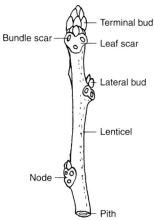

2

Identification Features

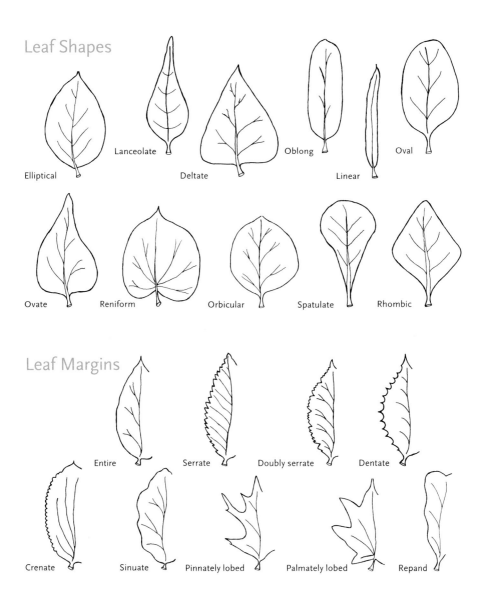

Leaf Shapes

Elliptical

Lanceolate

Deltate

Oblong

Linear

Oval

Ovate

Reniform

Orbicular

Spatulate

Rhombic

Leaf Margins

Entire

Serrate

Doubly serrate

Dentate

Crenate

Sinuate

Pinnately lobed

Palmately lobed

Repand

Leaf Apices

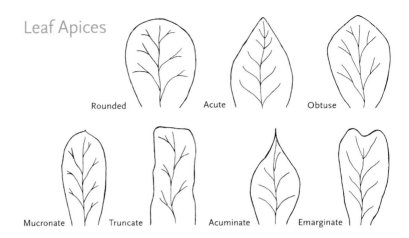

Rounded   Acute   Obtuse

Mucronate   Truncate   Acuminate   Emarginate

Leaf Bases

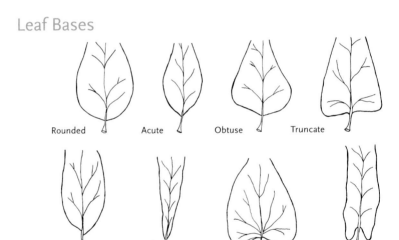

Rounded   Acute   Obtuse   Truncate

Inequilateral   Cuneate   Cordate   Auriculate

3
Guide to the Identification of Trees

Guide to the Gymnosperms

1. Leaves needlelike or linear, in bundles or solitary, two to four sided; cone woody or papery, with spiral scales. **Pine Family (*Pinaceae*)**

2. Leaves linear, awl-like, or scalelike, not in bundles; cone berry-like. **Cypress Family (*Cupressaceae*)**

3. Leaves linear, awl-like, or scalelike, not in bundles; cone round, with woody peltate scales. **Cypress Family (*Cupressaceae*)**

Guide to Cypress Family (*Cupressaceae*)

1. Leaves scalelike; cone round, about 6 mm (0.2 in) wide, with woody peltate scales. **Atlantic white-cedar (*Chamaecyparis thyoides*), see page 31**

2. Leaves scalelike, awl-like on seedlings; cone blue, waxy, and berrylike. **Eastern redcedar (*Juniperus virginiana* var. *virginiana*), see page 33**

3. Leaves linear and flat; cone up to 3.5 cm (1.4 in) wide, round, with woody peltate scales. **Baldcypress (*Taxodium distichum*), see page 37**

4. Leaves awl-like and overlapping; cone up to 3.5 cm (1.4 in) wide, round, with woody peltate scales. **Pondcypress (*Taxodium ascendens*), see page 35**

Guide to the Pine Family (*Pinaceae*)
I. Needles not in bundles.

1. Needles two-sided, somewhat two-ranked, white bands on the underside, born on pegs; cone small, 1.3–2.0 cm (0.5–0.8 in) long, pendent, lacking prickles; bark red-brown and deeply grooved. **Eastern hemlock (*Tsuga canadensis*), see page 55**

II. Needles only in bundles of two.

1. Needles not twisted, flexible, 6–13 cm (2.4–5.1 in) long; cone 4–7 cm (1.6–2.8 in) long, with small sharp prickles; bark plated, with resin holes. **Shortleaf pine (*Pinus echinata*), see page 39**

2. Needles sometimes twisted, flexible, 5–10 cm (2.0–3.8 in) long; cone 4–8 cm (1.6–3.1 in) long, with weak deciduous prickles; bark on large trees furrowed rather than plated. **Spruce pine (*Pinus glabra*), see page 43**

3. Needles twisted, stiff, 3–8 cm (1.2–3.1 in) long; cone 4–7 cm (1.6–2.8 in) long, with sharp prickles; bark orange-brown, scaly. **Virginia pine (*Pinus virginiana*), see page 53**

III. Needles in bundles of two and three.

1. Needles 15–28 cm (5.9–11.0 in) long; cone 7–18 cm (2.8–7.1 in) long, chocolate brown, and shiny, with weak prickles; bark plated, possibly tan flecked. **Slash pine (*Pinus elliottii*), see page 41**

IV. Needles in bundles of more than two.

1. Needles in bundles of five, blue-green, 8–14 cm (3.1–5.5 in) long; cone 8–20 cm (3.1–7.9 in) long, slim, and lacking prickles; branches in whorls. **Eastern white pine (*Pinus strobus*), see page 49**

2. Needles in bundles of three, 20–46 cm (7.9–18.1 in) long, clustered in tufts at the ends of stout branches; cone 15–25 cm (5.9–9.8 in) long, with sharp prickles; bark with red-brown plates. **Longleaf pine (*Pinus palustris*), see page 45**

3. Needles in bundles of three, 13–21 cm (5.1–8.3 in) long; cone 5–7 cm (2.0–2.8 in) long, serotinous; bark may show tufts of needles or sprouts. **Pond pine (*Pinus serotina*), see page 47**

4. Needles in bundles of three, 12–23 cm (4.7–9.1 in) long; cone 6–14 cm (2.4–5.5 in) long, gray-brown, with sharp prickles; bark with red-brown or brown plates. **Loblolly pine (*Pinus taeda*), see page 51**

Guide to the Angiosperms

I. Leaves opposite or whorled and simple.

1. Leaves opposite or whorled, heart shaped, margin entire; flowers white and trumpetlike; fruit long and beanlike; bark scaly or grooved. **Catalpa Family (***Bignoniaceae***), see page 91**

2. Leaves opposite, heart shaped, margin entire, underside tomentose; seedlings may show a shallow lobe or coarse tooth; flowers purple and trumpetlike; fruit a persistent nutlike capsule; bark smooth. **Figwort Family (***Scrophulariaceae***), see page 285**

3. Leaves opposite, margin serrate or entire, underside possibly glandular or rusty pubescent; flowers white, in flat-topped clusters; fruit a drupe. **Moscatel Family (***Adoxaceae***), see page 57**

4. Leaves opposite or whorled, margin entire, venation arcuate; flower clusters surrounded by petal-like bracts or white flat-topped heads lacking showy bracts; fruit a red or blue drupe. **Dogwood Family (***Cornaceae***)**

5. Leaves opposite, margin entire, aromatic when crushed, shiny upper surface and white underside; flowers dark red, smell like strawberries; fruit a leathery capsule. **Strawberry-Shrub Family (***Calycanthaceae***), see page 93**

6. Leaves opposite or subopposite, margin entire or serrate; leaf scar with one bundle scar; fruit a blue drupe. **Olive Family (***Oleaceae***)**

7. Leaves opposite, palmately lobed, margin can be serrate; fruit a double samara. **Soapberry Family (***Sapindaceae***)**

II. Leaves opposite and compound.

1. Leaves palmately compound with mostly five leaflets. **Soapberry Family (***Sapindaceae***)**

2. Leaves trifoliately compound. **Bladdernut Family (***Staphyleaceae***), see page 289**

3. Leaves pinnately compound, with three to nine leaflets, often shallowly and irregularly lobed; twigs bright green; fruit a double samara. **Soapberry Family (***Sapindaceae***)**

4. Leaves pinnately compound, with five or more leaflets, margin serrate or entire; fruit a single samara; buds with dark scales. **Olive Family (***Oleaceae***)**

III. Leaves alternate, simple, and lobed.

1. Leaves fan shaped, 13–20 cm (5.1–7.9 in) wide, with three to five lobes, margin coarsely toothed. **Sycamore Family (***Platanaceae***)**
2. Leaves tulip shaped with four broad lobes, apex broadly notched. **Magnolia Family (***Magnoliaceae***)**
3. Leaves star shaped. **Sweetgum Family (***Altingiaceae***), see page 59**
4. Leaves pinnately lobed; buds clustered at twig ends; fruit an acorn; pith star shaped. **Beech Family (***Fagaceae***)**
5. Leaves shallowly lobed at the apex; buds clustered at twig ends; fruit an acorn; pith star shaped. **Beech Family (***Fagaceae***)**

IV. Leaves alternate, simple, and unlobed: fragrant when crushed.

1. Leaves with an anise smell when crushed, leathery, with obscure lateral venation. **Star-Vine Family (***Schisandraceae***), see page 283**
2. Leaves yellow and glandular, with a spicy aroma when crushed; fruit a waxy drupe. **Wax-Myrtle Family (***Myricaceae***), see page 229**
3. Leaves unlobed or three-lobed, or with insect galls, with a spicy aroma when crushed. **Laurel Family (***Lauraceae***)**

V. Leaves alternate, simple, and unlobed: twigs armed.

1. Leaves with an acuminate apex, long petiole exuding milky sap when cut, margin entire; fruit a large, green brainlike ball of drupes. **Fig Family (***Moraceae***)**
2. Leaves with an entire margin; twigs with milky sap; fruit a black drupe. **Sapodilla Family (***Sapotaceae***)**
3. Leaf margin entire, serrate or irregularly toothed; twigs may have a bitter almond smell when cut; some species with a pair of glands on the petiole; flowers with five petals; fruit a drupe or pome. **Rose Family (***Rosaceae***)**

VI. Leaves alternate, simple, and unlobed: margins without teeth.

A. Leaves evergreen.

1. Leaves elliptical or oblong, leathery, sweet tasting, midrib bright yellow; flowers resemble yellow pom-poms; young bark streaked. **Sweetleaf Family (***Symplocaceae***), see page 297**

2. Leaves elliptical to oblong; twigs with stipule scars completely encircling the twig; flowers white and fragrant; fruit-red follicles in a conelike structure. **Magnolia Family (*Magnoliaceae*)**

B. Leaves deciduous.

i. Leaves heart shaped or triangular.

1. Leaves triangular with a pair of glands at the blade on a long petiole; fruit like a popcorn kernel, waxy. **Spurge Family (*Euphorbiaceae*), see page 117**

2. Leaves heart shaped, petiole swollen at both ends; flowers pink in early spring; fruit a legume. **Bean or Pea Family (*Fabaceae*)**

ii. Leaves neither heart shaped nor triangular.

1. Leaves obovate to oblong, with a green tomato smell when crushed; leaf underside, twig, and buds with velvety maroon hair; buds naked; flowers with maroon petals in threes; fruit banana-like. **Custard-Apple Family (*Annonaceae*), see page 69**

2. Leaves elliptical, ovate, obovate, oblong, or oval; twigs with stipule scars completely encircling the twig; flowers white or yellow; fruit red follicles in a conelike structure. **Magnolia Family (*Magnoliaceae*)**

3. Leaves lanceolate, elliptical, or oblong, with black splotches in late summer; twigs with orange lenticels; buds black and triangular; fruit a large edible berry; bark on older trees almost black and alligator-like. **Ebony Family (*Ebenaceae*), see page 111**

4. Leaves elliptical, oblong, ovate, oval, obovate, lanceolate, or oblanceolate; margin occasionally with a few coarse teeth near the apex; fruit a juicy purple-black drupe, with a ribbed or winged stone. **Dogwood Family (*Cornaceae*)**

5. Leaves elliptical, oblong, lanceolate, oblanceolate, obovate, or spatulate; leaves may show a bristle tip at the apex; buds clustered at twig ends; fruit an acorn; pith star shaped. **Beech Family (*Fagaceae*)**

6. Leaves obovate, with a long purple-red petiole; leaf scar bluish; fruit a drupe, on feathery stalks. **Cashew Family (*Anacardiaceae*)**

7. Leaves obovate to nearly round, margin may show irregular teeth; leaf scar with one bundle scar; flowers like white bells; fruit a wingless drupe. **Storax Family (*Styracaceae*)**

VII. Leaves alternate, simple, unlobed; margins with teeth.

A. Leaves tardily deciduous or evergreen.

1. Leaves tardily deciduous, oval to elliptical, glossy, margin with small glandular teeth; flowers bell shaped; fruit a blue berry. **Heath Family (*Ericaceae*)**

2. Leaves evergreen and leathery, margin with spiny teeth or an occasional bristle-tipped tooth or crenate; fruit a round shiny red to dark blue drupe; bark smooth even on large trees. **Holly Family (*Aquifoliaceae*)**

3. Leaves evergreen, elliptical, leathery, with small sharp teeth on the margin; flowers with five white, silky-haired petals; fruit a capsule. **Tea Family (*Theaceae*), see page 299**

4. Leaves evergreen, elliptical, leathery, margins with scattered hooked red teeth; fruit a blue-black drupe; bark smooth and gray. **Rose Family (*Rosaceae*)**

B. Leaves deciduous and heart shaped or triangular.

1. Leaves heart shaped, leaf margin coarsely serrate, petiole with milky sap when cut, upper surface possibly sandpapery, some species with one to three lobes; fruit a single drupe or an aggregate of drupes. **Fig Family (*Moraceae*)**

2. Leaves heart shaped, base inequilateral, margin with fine teeth; fruit a nutlet attached to an unlobed leafy bract. **Hibiscus or Mallow Family (*Malvaceae*), see page 221**

3. Leaves triangular to nearly round, margin with coarse rounded teeth, petiole long, with glands near the blade. **Willow or Poplar Family (*Salicaceae*)**

C. Leaves deciduous with singly serrate or irregularly toothed margins.

1. Leaves glossy, with prominent parallel venation, elliptical or oblong, margin finely serrate; fruit a red to black drupe; bark streaked. **Buckthorn Family (*Rhamnaceae*), see page 241**

2. Leaves obovate or elliptical, with sunken lateral veins, margin finely toothed, somewhat wavy; lateral bud stalked with two to three maroon valvate scales; fruit a nutlet in a woody cone. **Birch Family (*Betulaceae*)**

3. Leaves with three veins arising from the leaf base; base cordate

or rounded; fruit a berrylike drupe. **Cannabis and Hop Family (*Cannabaceae*)**

 4. Leaves elliptical to oval, margins serrate or irregularly toothed; some species with thorns or pair of glands on the petiole; twigs may have a bitter almond smell when cut; flowers with five petals; fruit a drupe or pome. **Rose Family (*Rosaceae*)**

 5. Leaves elliptical to ovate and nearly round, irregularly toothed or with a wavy margin; flowers like white bells; fruit a dry winged or dry wingless drupe; young bark may appear striped. **Storax Family (*Styracaceae*)**

 6. Leaves serrate or crenate, elliptical; fruit a round shiny red drupe; bark smooth even on large trees. **Holly Family (*Aquifoliaceae*)**

 7. Leaves with blunt teeth or upwardly curved teeth; teeth bristle tipped or callous tipped; fruit an acorn or a nut in a spiny bur. **Beech Family (*Fagaceae*)**

 8. Leaves finely serrate with gland-tipped teeth, lanceolate; buds with one scale; mature bark brown with loose plates. **Willow or Poplar Family (*Salicaceae*)**

 9. Leaves serrate with fine teeth, elliptical, midrib with long stiff hairs, sour tasting; fruit in persistent drooping clusters; mature bark dark and deeply grooved with orange in the grooves. **Heath Family (*Ericaceae*)**

D. Leaves deciduous with doubly serrate margins.

 1. Leaves doubly serrate, ovate or elliptical; fruit a nutlet in a papery cone, hoplike sac or three-lobed leafy bract. **Birch Family (*Betulaceae*)**

 2. Leaves doubly serrate and serrate, elliptical to ovate or obovate, base inequilateral or rounded; buds divergent; fruit a round or elliptical samara or burlike drupe. **Elm Family (*Ulmaceae*)**

VIII. Leaves alternate and compound: twigs armed.

A. Leaves pinnately compound.

 1. Leaves 20–30 cm (7.9–11.8 in) long, with 7 to 19 leaflets; leaflets 2–5 cm (0.8–2.0 in) long; twigs with a pair of spines at the node; flowers in dangling, fragrant, white clusters; fruit a legume. **Pea or Bean Family (*Fabaceae*)**

 2. Leaves up to 38 cm (15 in) long, with 5 to 19 leaflets; leaflets

2.5–7.0 cm (1.0–2.8 in) long, margin teeth with yellow glands; twigs with spines or prickles; trunk with pyramidal growths. **Rue Family (*Rutaceae*), see page 257**

B. Leaves bipinnately compound.

1. Leaves pinnately and bipinnately compound, up to 20 cm (7.9 in) long, with 14 to 30 leaflets; leaflets about 2 cm (0.8 in) long; twigs with branched thorns; fruit a long (up to 45 cm [17.7 in]), twisted legume; bark often with clusters of thorns. **Pea or Bean Family (*Fabaceae*)**

2. Leaves bi- and tripinnately compound, up to 1.6 m (5.2 ft) long, rachis and leaf scar spiny; twigs with spines or prickles; flowers in large drooping clusters; fruit a purple drupe in heavy clusters; usually single-stemmed. **Ginseng Family (*Araliaceae*), see page 79**

IX. Leaves alternate and compound; twigs unarmed.

A. Leaves pinnately compound.

1. Leaves up to 40 cm (15.7 in) long, with 7 to 13 leaflets; leaflets 5–10 cm (2.0–3.9 in) long, margins entire, rachis bright red; twigs stout with large shield-shaped leaf scars; fruit a persistent white drupe; young bark with black patches. Do not touch! **Cashew Family (*Anacardiaceae*)**

2. Leaves up to 1 m (3 ft) long, with 11 to 41 leaflets; leaflets basally lobed with black glands on the lobe underside; twigs stout and curving upward; fruit a single samara in persistent clusters; mature bark smooth. **Quassia Family (*Simaroubaceae*), see page 287**

3. Leaves up to 60 cm (23.6 in) long, with 8 to 22 leaflets on mature trees (may be fewer on seedlings), margin serrate, lemon scent when crushed; twigs with three-lobed leaf scars; fruit a nut in a yellow-green, semi-fleshy husk. **Walnut Family (*Juglandaceae*)**

4. Leaves up to 40 cm (15.7 in) long, with 9 to 31 leaflets, margin coarsely serrate or entire, rachis red or winged; buds fuzzy, somewhat encircled by the leaf scar; fruit a red drupe, conical, in upright terminal clusters. **Cashew Family (*Anacardiaceae*)**

B. Leaves bipinnately compound.

1. Leaves up to 45 cm (17.7 in) long, feathery; leaflets about 1 cm (0.4 in) long and narrow, margin entire; mature bark smooth;

flowers like pink pom-poms; fruit a legume. **Pea or Bean Family (*Fabaceae*)**

2. Leaves can also be tripinnately compound, up to 50 cm (19.7 in) long; leaflets 3–8 cm (1.2–3.1 in) long with a coarsely serrate margin, some basally lobed; flowers like small purple firecrackers; fruit a yellow drupe in persistent clusters; young bark purplish. **Mahogany Family (*Meliaceae*), see page 223**

Guide to the Cashew Family (*Anacardiaceae*)

1. Leaves simple, obovate to oval, petiole purple-red. **American smoketree (*Cotinus obovatus*), see page 61**

2. Leaves compound, with 7 to 13 leaflets; leaflets elliptical, 5–10 cm (2.0–3.9 in) long, elliptical, margin entire, rachis bright red; twig glabrous with large shield-shaped leaf scars. Do not touch! **Poison-sumac (*Toxicodendron vernix*), see page 67**

3. Leaves compound, with 9 to 21 leaflets; leaflets lanceolate or falcate, 4–7 cm (1.6–2.8 in) long, margin mostly entire, rachis green and winged; twig pubescent. **Winged sumac (*Rhus copallinum*), see page 63**

4. Leaves compound, with 11 to 31 leaflets; leaflets lanceolate or falcate, 5–14 cm (2.0–5.5 in) long, margin serrate, rachis red and glabrous; twig glaucous. **Smooth sumac (*Rhus glabra*), see page 65**

Guide to the Holly Family (*Aquifoliaceae*)

I. Leaves evergreen.

1. Leaves leathery, margin with prominent large spiny teeth; fruit a red drupe. **American holly (*Ilex opaca*), see page 75**

2. Leaves 4–9 cm (1.6–3.5 in) long, leathery, margin with bristle-tipped teeth above the middle; fruit a black drupe. **Large gallberry (*Ilex coriacea*), see page 71**

3. Leaves 1–4 cm (0.4–1.6 in) long, shiny, apex lacking a rigid spine, margin with small round teeth; fruit a red drupe. **Yaupon (*Ilex vomitoria*), see page 77**

II. Leaves deciduous.

1. Leaves elliptical, 4–8 cm (1.6–3.1 in) long, apex lacking a rigid spine, margin serrate to crenate; fruit a red drupe. **Possumhaw (*Ilex decidua*), see page 73**

Guide to the Birch Family (*Betulaceae*)

1. Leaves obovate, margin finely serrate and wavy, underside with maroon hairs; buds stalked; fruit a nutlet in a persistent woody cone. **Hazel alder (*Alnus serrulata*), see page 81**

2. Leaves triangular or ovate, base wedge shaped, margin doubly serrate; fruit a nutlet in a papery cone; young bark orange-brown to salmon-pink and peeling. **River birch (*Betula nigra*), see page 85**

3. Leaves ovate, base cordate or inequilateral, margin serrate or doubly serrate; twigs with a strong wintergreen odor when cut; fruit a nutlet in a papery cone; bark maroon-black with horizontal lenticels on young trees and with gray to black scaly plates on large trees. **Black birch (*Betula lenta*), see page 83**

4. Leaves ovate to elliptical, base cordate or inequilateral, leaf margin doubly serrate, lateral veins dividing near the leaf margin; buds green and brown striped; fruit a nutlet in a hoplike sac; bark shredding in thin strips. **Hophornbeam (*Ostrya virginiana*), see page 89**

5. Leaves ovate to elliptical, base cordate or inequilateral, margin doubly serrate; buds maroon and white striped; fruit a nutlet in a three-lobed leafy bract; bark smooth, gray, fluted. **Hornbeam (*Carpinus caroliniana*), see page 87**

Guide to the Cannabis and Hop Family (*Cannabaceae*)

1. Leaves ovate, apex acuminate, margin entire or irregularly serrate near the apex, three veins arising from the leaf base; twigs and buds pubescent; fruit an orange-red to purple drupe; bark smooth with corky warts. **Sugarberry (*Celtis laevigata*), see page 95**

2. Leaves ovate, apex acute or acuminate, margin serrate, three veins arising from the leaf base; twigs and buds mostly glabrous; fruit a red-purple drupe; bark with warty ridges. **Hackberry (*Celtis occidentalis*), see page 97**

3. Leaves ovate, apex acute or acuminate, margin irregularly serrate, three veins arising from the leaf base, surface scabrous; fruit an orange-red drupe; bark with corky warts. **Georgia hackberry (*Celtis tenuifolia*), see page 99**

Guide to the Dogwood Family (*Cornaceae*)

I. Leaves opposite.

1. Leaves clearly opposite, with arcuate venation; flower clusters

surrounded by four petal-like white bracts; fruit a shiny red drupe in erect clusters; bark blocky. **Flowering dogwood (*Cornus florida*), see page 103**

II. Leaves alternate.

1. Leaves may appear opposite or whorled, petiole long, venation arcuate; flowers in white flat-topped clusters, lacking showy petal-like bracts; fruit a purple-blue drupe on bright red stalks; young bark green with white streaks. **Alternate-leaf dogwood (*Cornus alternifolia*), see page 101**

2. Leaves up to 30 cm (11.8 in) long; fruit a drupe, long stalked, large (3 cm [1.2 in] long), and juicy, with a longitudinally ridged stone. **Water tupelo (*Nyssa aquatica*), see page 105**

3. Leaves obovate to oval, up to 16 cm (6.3 in) long; fruit a drupe, 1.3 cm (0.5 in) long, in groups of three to five, stone shallowly ribbed; mature bark typically blockier and thicker than that of swamp tupelo (*Nyssa biflora*). **Blackgum (*Nyssa sylvatica*), see page 109**

4. Leaves up to 15 cm (5.9 in) long, elliptical or lanceolate; fruit a drupe, 1.0–1.5 cm (0.4–0.6 in) long, grows in pairs, stone ribbed; bark shallowly grooved; habitat usually wetter than for blackgum (*Nyssa sylvatica*). **Swamp tupelo (*Nyssa biflora*), see page 107**

Guide to the Bean or Pea Family (*Fabaceae*)

1. Leaves simple, heart shaped, margin entire, petiole swollen at both ends; twigs dark and zigzag; flowers in pink showy clusters in early spring; fruit podlike, up to 10 cm (3.9 in) long; bark red-brown to black, shallowly ridged or scaly. **Eastern redbud (*Cercis canadensis*), see page 121**

2. Leaves pinnately or bipinnately compound, up to 20 cm (7.9 in) long, with 14 to 30 leaflets; leaflets 2 cm (0.8 in) long; twigs with branched thorns; fruit podlike, up to 45 cm (17.7 in) long, often twisted; bark may have clumps of thorns. **Honeylocust (*Gleditsia triacanthos*), see page 123**

3. Leaves pinnately compound, 20–36 cm (7.9–14.1 in) long, with 7 to 19 leaflets; leaflets 2–5 cm (0.8–2.0 in) long; twigs with a stout pair of spines at each node, especially on younger twigs or sprouts; fruit podlike, up to 10 cm (3.9 in) long; bark furrowed with interlacing fibrous ridges. **Black locust (*Robinia pseudoacacia*), see page 125**

4. Leaves bipinnately compound, up to 45 cm (17.7 in) long; leaflets about 1 cm (0.4 in) long and lopsided; flowers resembling pink pom-poms; fruit podlike, up to 20 cm (7.9 in) long; bark smooth and gray. **Mimosa (*Albizia julibrissin*), see page 119**

Guide to the Beech Family (*Fagaceae*)

I. Fruit is a nut in a spiny bur; branch terminal with a single bud.

1. Leaf 13–20 cm (5.1–7.9 in) long, margin with curved, coarse bristle-tipped teeth; leaf underside, twigs, and buds glabrous; two to three nuts enclosed in a spiny bur. **American chestnut (*Castanea dentata*), see page 127**

2. Leaf 8–15 cm (3.1–5.9 in) long, margin with coarse bristle-tipped teeth; leaf underside and twigs with tomentose hair; one nut enclosed in a spiny bur. **Allegheny chinkapin (*Castanea pumila*), see page 129**

3. Leaf 8–13 cm (3.1–5.1 in) long, margin serrate, parallel lateral veins ending at a tooth; buds long and sharp pointed (spear shaped); two small triangular nuts in a weakly spiny bur; bark gray and smooth, even on large trees. **American beech (*Fagus grandifolia*), see page 131**

II. Fruit an acorn; branch terminal with a cluster of buds.

A. Lobed red oaks: lobes bristle tipped.

1. Leaves shallowly three-lobed at the apex, fan shaped, leathery, underside with orange-brown hair; acorn 2 cm (0.8 in) long, cap covering up to one-half of the nut, nut with a rigid point; bark dark, rough, blocky; habitat dry sites. **Blackjack oak (*Quercus marilandica*), see page 155**

2. Leaves with three to seven lobes, terminal lobe elongated, sinuses deep, base bell shaped, underside densely hairy; acorn 1.3 cm (0.5 in) long, cap covering up to one-half of the nut; bark dark and rough; habitat dry sites. **Southern red oak (*Quercus falcata*), see page 141**

3. Leaves with three to seven lobes, central lobe elongated, sinuses deep, base tapered, pubescence only in vein axils; acorn 2 cm (0.8 in) long, cap with a rolled edge and covering one-third of the nut; habitat sandy soils. **Turkey oak (*Quercus laevis*), see page 147**

4. Leaves with five to seven lobes, lobing irregular on the same leaf, terminal lobe elongated, sinuses deep; acorn 3 cm (1.2 in) long, cap covering up to one-half of the striped nut; bark smooth or shallowly grooved; habitat bottomlands. **Nuttall oak (***Quercus texana***), see page 175**

5. Leaves with five to seven lobes, lobing and sinus depth variable from lower to upper canopy leaves, leaf underside and end of twig often pubescent; bud with overlapping rows of pubescent scales; acorn 1–2 cm (0.4–0.8 in) long, cap slightly fringed at base and covering up to one-half of the nut; bark dark and ridged or blocky; habitat uplands. **Black oak (***Quercus velutina***), see page 177**

6. Leaves with five to seven lobes, basal lobes wide, sun leaves like southern red oak (*Quercus falcata*) except not bell shaped at the base, underside densely pubescent; acorn 1.3 cm (0.5 in) long, the cap covering up to one-half of the nut; bark with flaky or scaly ridges; habitat bottomlands. **Cherrybark oak (***Quercus pagoda***), see page 165**

7. Leaves with five to nine lobes, lobes with many bristle tips, sinuses deep, underside glabrous or with tufts of hair in vein axils; acorn 2–3 cm (0.8–1.2 in) long, cap with rolled edges covering up to one-third of the striped nut; bark smooth or shallowly fissured; habitat bottomlands. **Shumard oak (***Quercus shumardii***), see page 171**

8. Leaves with five to nine lobes, sinuses deep, underside glabrous or with tufts of hair in vein axils; acorn up to 2.5 cm (1 in) long, cap covering up to one-half of the nut, nut with concentric grooves at the tip; bark dark with rough and possibly white-streaked ridges; habitat dry uplands. **Scarlet oak (***Quercus coccinea***), see page 137**

9. Leaves with seven to eleven lobes, sinuses less than halfway to the midrib, underside glabrous or with tufts of hair in vein axils; acorn 2–3 cm (0.8–1.2 in) long, cap with rolled edges covering at most one-fourth of the nut; bark with flattened gray or white ridges; habitat moist uplands. **Northern red oak (***Quercus rubra***), see page 169**

B. Unlobed red oaks: apex with a bristle tip.

1. Leaves oblong, 5–12 cm (2.0–4.7 in) long, margin mostly entire, underside with dense blue-white pubescence; acorn about 1 cm (0.4 in) long, cap covering up to one-third of the nut; habitat sandy soils. **Bluejack oak (***Quercus incana***), see page 145**

2. Leaves oblanceolate to elliptical, 5–12 cm (2.0–4.7 in) long,

shiny, margin mostly entire, underside glabrous; acorn about 1.3 (0.5 in) cm long, cap covering one-fourth or less of the nut; habitat uplands. **Laurel oak (*Quercus hemisphaerica*), see page 143**

3. Leaves spatulate, obovate, oblanceolate, subrhombic, 5–14 cm (2.0–5.5 in) long, margin mostly entire, underside mostly glabrous; acorn 1.5 cm (0.6 in) long, cap covering one-fourth to one-half of the nut; habitat bottomlands and swamps. **Swamp laurel oak (*Quercus laurifolia*), see page 149**

4. Leaves linear to lanceolate, 5–12 cm (2.0–4.7 in) long, thin, margin entire, midrib with yellow pubescence possible; acorn up to 1.3 cm (0.5 in) long, cap covering one-fourth or less of the nut; habitat bottomlands. **Willow oak (*Quercus phellos*), see page 167**

5. Leaves occasionally lobed (especially on young trees), 5–10 cm (2.0 3.9 in) long, margin mostly entire, underside mostly glabrous; acorn 1.2 cm (0.5 in) long, cap covering one-fourth of the nut; habitat uplands to bottomlands. **Water oak (*Quercus nigra*), see page 163**

C. Lobed white oaks: lobes and apex lacking bristle tips.

1. Leaves 5–15 cm (2.0–5.9 in) long, with three to seven irregular lobes, underside glabrous; acorn 2 cm (0.8 in) long, cap covering one-fourth to one-third of the nut; habitat river bluffs. **Bluff oak (*Quercus austrina*), see page 135**

2. Leaves 10–15 cm (3.9–5.9 in) long, with five variable lobes, some central lobes cruciform, underside pubescent to tomentose; acorn 1.3–2.0 cm (0.5–0.8 in) long, cap covering one-third to one-half of the nut; habitat well-drained soils. **Post oak (*Quercus stellata*), see page 173**

3. Leaves 2–15 cm (0.8–5.9 in) long, with three to five variable lobes, usually pointing toward the apex, central lobes only slightly cruciform, underside pubescent; acorn 1.3–2.0 cm (0.5–0.8 in) long, cap covering one-third to one-half of the nut; habitat dry soils. **Sand post oak (*Quercus margarettiae*), see page 153**

4. Leaves 12–20 cm (4.7–7.9 in) long, with five to nine lobes, lobes and sinuses very irregular, underside pubescent; acorn 2 cm (0.8 in) long, cap almost completely covering the nut; habitat bottomlands. **Overcup oak (*Quercus lyrata*), see page 151**

5. Leaves 10–18 cm (3.9–7.1 in) long, with seven to nine lobes, lobing more regular, sinuses extending halfway or more to the midrib, underside pale and glabrous; acorn 2–3 cm (0.8–1.2 in) long, cap

covering one-fourth to one-third of the nut; habitat well-drained soils. **White oak (*Quercus alba*), see page 133**

D. Unlobed white oaks: leaves of most species lacking bristles at the apex and margin, some with scalloped margins.

1. Leaves oblong to oblanceolate, 4–15 cm (1.6–5.9 in) long, very leathery, apex with or without a rigid bristle tip, leaf margin entire or with an occasional bristle, underside pubescent; acorn 1.0–2.5 cm (0.4–1.0 in) long, cap long stalked and covering up to one-half of the nut; habitat sandy soils. **Live oak (*Quercus virginiana*), see page 179**

2. Leaves obovate to spatulate, 12–18 cm (4.7–7.1 in) long, leaf margin sinuate, petiole yellow, underside with dense pubescence; acorn 1.3–2.0 cm (0.5–0.8 in) long, cap saucer shaped and covering only the base of the nut; habitat limestone soils and bottomlands. **Durand oak (*Quercus durandii*), see page 139**

3. Leaves obovate throughout the canopy, 12–22 cm (4.7–8.7 in) long, margin scalloped with rounded callous-tipped teeth, underside pubescent; acorn 2.5–4.0 cm (1.0–1.6 in) long, cap bowl shaped with a thin edge covering one-third of the nut; habitat bottomlands. **Swamp chestnut oak (*Quercus michauxii*), see page 157**

4. Leaf elliptical in the upper canopy and obovate in shade leaves, 9–18 cm (3.5–4.7 in) long, leaf margin scalloped with curved callous tipped teeth, underside pubescent; acorn 1.3–2.5 cm (0.5–1.0 in) long, nut brown-black, cap with a thin edge covering one-third of the nut; habitat alkaline soils. **Chinkapin oak (*Quercus muehlenbergii*), see page 161**

5. Leaves obovate to elliptical, 10–22 cm (3.9–8.7 in) long, leaf margin scalloped with rounded smooth teeth; acorn 2–4 cm (0.8–1.6 in) long, nut yellow-brown when young, cap with fused warty scales covering one-third of the nut and becoming loose on the nut; bark deeply grooved (swamp chestnut oak [*Quercus michauxii*] and chinkapin oak have scaly bark [*Quercus michauxii*]); habitat uplands. **Chestnut oak (*Quercus montana*), see page 159**

Guide to the Walnut Family (*Juglandaceae*)

I. Leaves with 8 to 24 leaflets; twig pith chambered; nut corrugated or rugose, enclosed in an indehiscent husk.

1. Leaves with 11 to 17 leaflets; leaflets ovate, with a terminal leaflet, rachis very hairy; leaf scar with a fuzzy mustache; nut elongated,

about 6 cm (2.4 in) long; husk sticky-pubescent; bark ash-gray with pale flattened ridges. **Butternut (*Juglans cinerea*), see page 201**

2. Leaves with 8 to 24 leaflets; leaflets ovate, often lacking the terminal leaflet; nut round, up to 6 cm (2.4 in) wide; husk yellow-green and somewhat fleshy; bark dark brown and furrowed with interlacing ridges. **Black walnut (*Juglans nigra*), see page 203**

II. Nut smooth or ribbed, enclosed in a dehiscent husk; pith solid.

A. Leaves with five to nine leaflets; buds imbricate; fruit husk unwinged at sutures.

1. Leaves with five to seven leaflets; leaflets obovate to lanceolate, 8–15 cm (3.1–5.9 in) long, underside mostly glabrous, rachis glabrous; twigs thin and glabrous; nut 2.5–5.0 cm (1–2 in) long; husk thin, shiny, round or pear shaped, and splitting halfway to the base; bark with ridges forming a diamond pattern. **Pignut hickory (*Carya glabra*), see page 185**

2. Similar to pignut hickory (*Carya glabra*) but petiole often red; nut husk rough and splitting to the base; bark with scaly ridges. **Red hickory (*Carya ovalis*), see page 193**

3. Leaves with five to seven leaflets; leaflets mostly obovate, 8–18 cm (3.1–7.1 in) long, with minute tufts of hair on margin teeth; nut up to 4 cm (1.6 in) long, round, with four ridges; husk thick (1.2 cm [0.5 in]), with four deep sutures that split to the base; bark shaggy with curved loose plates. **Shagbark hickory (*Carya ovata*), see page 195**

4. Similar to shagbark hickory (*Carya ovata*) but with seven larger (5–22 cm [2.0–8.7 in] long) leaflets; nut up to 6 cm (2.4 in) long with four to six ridges; husk thick (1.2 cm [0.5 in]); twig orange-brown or yellow. **Shellbark hickory (*Carya laciniosa*), see page 189**

5. Leaves with five to nine leaflets; leaflets 8–14 cm (3.1–5.5 in) long, underside with silvery scales or pale, rachis pubescent; bud with yellow scales; nut up to 3.5 cm (1.4 in) wide; husk with yellow scales and splitting to the base; bark with interlacing ridges. **Sand hickory (*Carya pallida*), see page 197**

6. Leaves with seven to nine leaflets; leaflets 5–18 cm (2.0–7.1 in) long; rachis and underside densely pubescent; twigs thick, with hair; nut up to 5 cm (2 in) long, nearly round; husk thick, splits to the base; bark with interlacing ridges forming a diamond pattern. **Mockernut hickory (*Carya tomentosa*), see page 199**

B. Leaves with nine (sometimes seven) or more leaflets; buds valvate; fruit husk winged at sutures.

   1. Leaves with 7 to 11 leaflets; leaflets obovate to lanceolate, 5–15 cm (2.0–5.9 in) long; buds sulfur-yellow and scurfy; nut nearly round and about 2.5 cm (1 in) long; husk yellow and scurfy with winged sutures above the middle; bark with tight interlacing ridges. **Bitternut hickory (*Carya cordiformis*), see page 183**

   2. Leaves with mostly nine leaflets; leaflets 5–15 cm (2.0–5.9 in) long, with yellow or silver scales on the underside or possibly shiny; twigs yellow-scurfy; buds yellow and scurfy; nut ellipsoidal and up to 4 cm (1.6 in) long; husk with winged sutures; bark shaggy with loose plates. **Nutmeg hickory (*Carya myristiciformis*), see page 191**

   3. Leaves with 7 to 13 leaflets; leaflets 5–12 cm (2.0–4.7 in) long, falcate, underside rusty colored with scales and pubescence; buds yellow-brown and scurfy; nut flattened, 2.5–4.0 cm (1.0–1.6 in) long; husk with winged sutures; bark scaly or shaggy. **Water hickory (*Carya aquatica*), see page 181**

   4. Leaves with 9 to 17 leaflets; leaflets 8–15 cm (3.1–5.9 in) long, falcate, underside with pale pubescence; buds gray-brown to yellowish and pubescent; nut ellipsoidal and 2.5–5.0 cm (1–2 in) long; husk with thinly winged sutures; bark silver-gray and scaly. **Pecan (*Carya illinoinensis*), see page 187**

Guide to the Laurel Family (*Lauraceae*)

   1. Leaves evergreen, unlobed, often disfigured, underside pale, fragrant when crushed; fruit a blue drupe on green stalks. **Redbay (*Persea borbonia*), see page 205**

   2. Leaves deciduous, unlobed or with two or three lobes, prominent venation, fragrant when crushed; fruit a blue drupe on bright-red stalks. **Sassafras (*Sassafras albidum*), see page 207**

Guide to the Magnolia Family (*Magnoliaceae*)

I. Leaves evergreen or semi-evergreen.

   1. Leaves semi-evergreen or evergreen, 8–17 cm (3.1–6.7 in) long, underside prominently silvery blue; flower up to 8 cm (3.1 in) wide; bark mostly smooth. **Sweetbay magnolia (*Magnolia virginiana*), see page 219**

   2. Leaves evergreen, 13–25 cm (5.1–9.8 in) long, underside rusty pubescent; flower up to 20 cm (7.9 in) wide, very fragrant; bark

smooth or flaky. **Southern magnolia (*Magnolia grandiflora*), see page 213**

II. Leaves deciduous.

1. Leaves 13–25 cm (5.1–9.8 in) long, usually with four lobes, apex broadly notched; fruit a cone of winged samaras; terminal bud valvate, up to 2 cm (0.8 in) long, flattened, with two scales; bark ash-gray and deeply furrowed. **Tulip-poplar (*Liriodendron tulipifera*), see page 209**

2. Leaves elliptical, 13–25 cm (5.1–9.8 in) long, apex acuminate, margin entire, underside silky pubescent; terminal bud 1.5 cm (0.5 in) long, with one pubescent scale; fruit a cone of red follicles, often curved like a cucumber; bark scaly. **Cucumbertree (*Magnolia acuminata*), see page 211**

3. Leaves obovate, 50–80 cm (19.7–31.5 in) long, earlike at the base, margin entire, underside with silver-white pubescence; terminal bud up to 6 cm (2.4 in) long, with one pubescent scale; fruit a cone of red follicles; bark smooth. **Bigleaf magnolia (*Magnolia macrophylla*), see page 215**

4. Leaves obovate, 25–60 cm (9.8–23.6 in) long, base acute, margin entire, underside with white pubescence, displayed at branch ends like an umbrella; terminal bud up to 5 cm (2 in) long, with one glabrous scale; fruit a cone of red follicles; bark mostly smooth. **Umbrella magnolia (*Magnolia tripetala*), see page 217**

Guide to the Fig Family (*Moraceae*)

1. Leaves ovate, 8–15 cm (3.1–5.9 in) long, margin entire; twigs often with a stout spine at the node; fruit a large green ball of drupes; inner bark orange. **Osage-orange (*Maclura pomifera*), see page 225**

2. Leaves ovate or orbicular, unlobed or lobed, 10–25 cm (3.9–5.9 in) long, margin serrate, upper surface scabrous; twigs without spines; fruit red-purple, blackberry-like; bark gray-brown. **Red mulberry (*Morus rubra*), see page 227**

Guide to the Olive Family (*Oleaceae*)

I. Leaves simple.

1. Leaves opposite, evergreen, more than 6 cm (2.4 in) long,

leathery; bark mottled and smooth. **Devilwood (*Osmanthus americanus*), see page 237**

2. Leaves opposite or subopposite, deciduous, more than 10 cm (5.9 in) long, petiole purplish; flowers four-petalled in fringelike clusters; bark grooved or scaly. **Fringe-tree (*Chionanthus virginicus*), see page 231**

II. Leaves pinnately compound.

1. Leaves with seven to nine leaflets; leaf scar shield shaped with the lateral bud above the scar; samara seed elongated, wing narrow and extending halfway down the seed. **Green ash (*Fraxinus pennsylvanica*), see page 235**

2. Leaves with five to nine leaflets; leaf scar crescent shaped with the lateral bud sitting within the scar; samara seed more rounded than for green ash (*Fraxinus pennsylvanica*), with the wing extending only to the top of the seed. **White ash (*Fraxinus americana*), see page 233**

Guide to the Rose Family (*Rosaceae*)

I. Leaves evergreen.

1. Leaves elliptical, 5–12 cm (2.0–9.8 in) long, margin entire or with an occasional hooked red tooth; flowers in short-stalked white racemes after the leaves; fruit a blue-black drupe, persisting over winter; bark gray-brown and smooth. **Cherry laurel (*Prunus caroliniana*), see page 251**

II. Leaves deciduous.

1. Leaves ovate, 4–10 cm (1.6–3.9 in) long, base cordate, margin serrate, pubescent when unfolding, petiole pubescent; buds long-pointed; flowers in white racemes before the leaves; fruit a red or purple pome; bark smooth and gray with vertical stripes or cracks. **Downy serviceberry (*Amelanchier arborea*), see page 243**

2. Leaves elliptical, 2.5–8.0 cm (1.0–3.1 in) long, margin finely serrate, base sometimes lobed, petiole pubescent; twigs with thorns; flowers in showy, fragrant, white-pink terminal racemes with the leaves; fruit a sour pome; bark scaly. **Southern crabapple (*Malus angustifolia*), see page 247**

3. Leaves lanceolate, 3–8 cm (1.2–3.1 in) long, often folding

upward, margin with glandular teeth, petiole with glands; twigs with thorns; flowers in white umbels before the leaves; fruit a yellow-red sweet drupe; young bark red-black and shiny. **Chickasaw plum (*Prunus angustifolia*), see page 249**

4. Leaves oblong-lanceolate or oval, 5–15 cm (2.0–5.9 in) long, margin serrate, midrib with tawny pubescence, petiole with one or two glands near the blade; flowers in white racemes after the leaves; fruit a purple-black, juicy drupe; mature bark with shiny black scales. **Black cherry (*Prunus serotina*), see page 255**

5. Leaves ovate, 5–13 cm (2.9–5.1 in) long, margin sharply serrate, underside densely pubescent, petiole pubescent and with or without glands; twigs pubescent; flowers in white umbels; fruit a large red drupe; bark scaly. **Mexican plum (*Prunus mexicana*), see page 253**

6. Leaves lobed, margin serrate or dentate; twigs with thorns; flowers white or pink, with five petals; fruit a round pome; bark smooth on small trees, scaly or peeling on large trees. **Hawthorn (*Crataegus* spp.), see page 245**

Guide to the Willow or Poplar Family (*Salicaceae*)

1. Leaves triangular, blade up to 18 cm (7.1 in) long, margin with coarse rounded teeth, petiole long and flattened; bark gray and deeply grooved. **Eastern cottonwood (*Populus deltoides*), see page 259**

2. Leaves lanceolate, 8–15 cm (3.1–5.9 in) long, margin with small red-tipped teeth, nearly sessile; bark brown with scaly ridges or loose plates. **Black willow (*Salix nigra*), see page 261**

Guide to the Soapberry Family (*Sapindaceae*)

I. Leaves palmately compound.

1. Twigs with an unpleasant odor when crushed; flowers with stamens longer than the petals; fruit husk prickly. **Ohio buckeye (*Aesculus glabra*), see page 265**

2. Twigs lacking odor; flowers with petals longer than stamens; fruit husk smooth. **Yellow buckeye (*Aesculus flava*), see page 263**

II. Leaves pinnately compound.

1. Leaves with three to seven leaflets; leaflets with coarse teeth and shallow lobes on the margins; twigs bright green with

V-shaped leaf scars; lateral buds yellow-white and pubescent. **Box-elder (*Acer negundo*), see page 271**

III. Leaves simple.

A. Margins not toothed.

1. Leaves with five palmate lobes, apices acuminate, underside pale and glabrous. **Sugar maple (*Acer saccharum*), see page 277**

2. Leaves with three to five palmate lobes, apices acute or blunt, underside white and usually pubescent. **Florida maple (*Acer florida-num*), see page 267**

3. Leaves palmately three- to five-lobed, apices acuminate, underside yellow-green, with or without pubescence. **Chalk maple (*Acer leucoderme*), see page 269**

B. Margins serrate.

1. Leaves with five palmate lobes, apices acuminate, sinuses V-shaped, margin coarsely serrate, underside silver-white. **Silver maple (*Acer saccharinum*), see page 275**

2. Leaves with three to five palmate lobes, apices acute, sinuses acute and shallow, margin irregularly doubly serrate. **Red maple (*Acer rubrum* var. *rubrum*), see page 273**

Guide to the Sapodilla Family (*Sapotaceae*)

1. Leaves 2–9 cm (0.8–3.5 in) long, underside with rusty or gray dense pubescence; twigs with rusty or gray pubescence. **Gum bumelia (*Sideroxylon lanuginosum*), see page 279**

2. Leaves 5–15 cm (2.0–5.9 in) long, underside mostly glabrous when mature; twigs glabrous. **Buckthorn bumelia (*Sideroxylon lycioides*), see page 281**

Guide to the Storax Family (*Styracaceae*)

I. Leaf margin regularly toothed.

1. Leaf elliptical, ovate or obovate, 5–15 cm (2.0–5.9 in) long, leaf margin with small sharp teeth; flower petals fused; fruit a four-winged drupe. **Carolina silverbell (*Halesia tetraptera*), see page 291**

II. Leaf margin irregularly toothed.

1. Leaf obovate or elliptical, 3–8 cm (1.2–3.1 in) long, leaf margin

irregularly toothed or wavy; flower petals fused only at the base and recurved; fruit an unwinged and pubescent drupe. **American snowbell (*Styrax americanus*), see page 293**

2. Leaf oval to orbicular, 8–16 cm (3.1–6.3 in) long, leaf margin with irregular teeth or entire; flower petals fused only at the base and not recurved; fruit an unwinged and pubescent drupe. **Bigleaf snowbell (*Styrax grandifolius*), see page 295**

Guide to the Elm Family (*Ulmaceae*)

1. Leaves two-ranked, ovate to lanceolate, 3–7 cm (1.2–2.8 in) long, apex acute, margin serrate; fruit with fleshy projections; bark scaly or shredding with red inner bark. **Water-elm (*Planera aquatica*), see page 301**

2. Leaves elliptical or lanceolate, 3–8 cm (1.2–3.1 in) long, apex acute, margin doubly serrate; twigs with corky wings, but not always; fruit an elliptical samara, apex notched, margin pubescent; bark with oblong scales. **Winged elm (*Ulmus alata*), see page 303**

3. Leaves obovate to elliptical, 8–15 cm (3.1–5.9 in) long, base greatly inequilateral, margin doubly serrate; twigs and buds mostly glabrous; fruit a round samara, apex notched, margin pubescent; bark with scaly, flattened ridges, and brown and white inner layers. **American elm (*Ulmus americana*), see page 305**

4. Leaves obovate to elliptical, 10–18 cm (3.9–7.1 in) long, margin doubly serrate, upper surface and underside scabrous; buds with maroon pubescence; fruit a round samara, apex shallowly notched, margin glabrous; bark red-brown with corky or flattened ridges and brown inner layers. **Slippery elm (*Ulmus rubra*), see page 307**

4
Species Accounts

Cypress Family (*Cupressaceae*)

Atlantic White-Cedar

Chamaecyparis thyoides (L.) Britton,
Sterns & Poggenb.

COMMON NAMES Atlantic white-cedar, southern
white-cedar, swamp-cedar

QUICK GUIDE Leaves scalelike in flat sprays; cone small and nearly
round with shield-shaped scales; bark ash-gray with fibrous interlacing ridges, often twisting around the trunk.

DESCRIPTION Leaves are up to 5 mm (0.2 in) long, scalelike, green to
blue-green, and displayed in flattened sprays; underside glandular.
The seed cone is nearly round, about 6 mm (0.2 in) wide, and ridged;
scales of mature cones are brown, woody, and peltate (shield shaped);
seed cones mature in one growing season. Bark is ash-gray with red-brown inner bark, with fibrous often stringy, interlacing ridges that

From left to right:

Atlantic white-cedar leaves.

Atlantic white-cedar seed cones.

From left to right:

Bark on young Atlantic white-cedar.

Bark on large Atlantic white-cedar.

may twist around the trunk, especially on large trees. The growth form is up to 24 m (80 ft) in height and 0.6 m (2 ft) in diameter, but old trees can be larger.

HABITAT Acidic freshwater bogs and swamps.

NOTES Often found growing with red maple, swamp cyrilla, southern magnolia, sweetbay magnolia, swamp tupelo, live oak, baldcypress, and pondcypress. Trees reach maturity in 150 to 200 years, but old-growth trees as old as 1,000 years have been reported. Most old-growth and pure stands have been cut. The wood is pale to pinkish in color, soft, straight-grained, light, and decay resistant. The wood is used for siding, shingles, boxes, fencing, small boat construction, pilings, poles, and water tanks.

Chamaecyparis means "low-growing cypress," perhaps referring to its low-lying habitat; *thyoides* means "similar to *Thuja*," referring to the leaves. White-cedar is hyphenated because it is not a true cedar.

Eastern Redcedar

Juniperus virginiana (L.) var. *virginiana*

COMMON NAMES eastern redcedar, red juniper

QUICK GUIDE Leaves scalelike or awl-like; cone smooth, berrylike, waxy blue; bark red-brown, fibrous, and often peeling.

DESCRIPTION Leaves are either awl-like (sharp looking) and about 8 mm (0.3 in) long or scalelike and up to 2 mm (0.1 in) long, green to yellow-green, and glandular. Awl-like leaves are typical on seedlings and saplings. The seed cone is about 5 mm (0.2 in) wide, round, berrylike, and blue with a white waxy coat, and matures in one growing season. This species is dioecious, so seed and pollen cones are on separate trees. Bark is red-brown to gray, peeling into long and fibrous or shreddy strips. The growth form is up to 15 m (50 ft) in height and 0.6 m (2 ft) in diameter.

HABITAT A wide variety of sites such as abandoned fields, swamp edges, and upland forests; fence rows in the Alabama Black Belt.

NOTES Eastern redcedar is found throughout Alabama and is the most widely distributed conifer in the eastern United States. Eastern

Eastern redcedar form and habitat.

redcedar is drought resistant and can tolerate thin rocky soils. Forest associates include many hickory species, common persimmon, sweetgum, numerous pines, winged elm, and a variety of oaks. The wood consists of yellow-white sapwood and red-maroon heartwood, and is fragrant and durable with many uses, including fence posts, railroad ties, furniture, paneling, cedar chests, pencils, woodenware, and cedarwood oil. Eastern redcedar is used in Christmas tree production and in windbreaks. The foliage is browsed by white-tailed deer, and the seed cones are eaten by a wide variety of game birds and songbirds as well as small to midsize mammals such as opossum and raccoon. The dense foliage provides excellent nesting, roosting, and foraging substrate for birds. Older trees are interesting because of their bark and low, wide, evergreen crown. Cultivars with better form and foliage color have been developed for landscaping. However, this species is a host for apple-rust disease, and galls may be visible on the branches.

Juniperus is Latin for "juniper tree"; *virginiana* refers to the geographic range. Redcedar is one word as it is not a true cedar.

SIMILAR SPECIES **Southern redcedar** (*Juniperus virginiana* var. *silicola* [Small] E. Murra) is similar and found on coastal sites (dunes and sandy ridges) and on brackish sites and is reported to have smaller cones and more drooping needlelike foliage.

Pondcypress

Taxodium ascendens Brongn.

COMMON NAMES pondcypress, black cypress

QUICK GUIDE Distinguished from baldcypress by appressed leaves and upwardly curving branchlets.

DESCRIPTION Leaves are deciduous, awl-like or linear, appressed, and green to yellow-green, turning red-brown in the autumn. From a distance the foliage may look like pine needles. Branchlets curve upward (ascend) and can be deciduous. The seed cone is up to 2.5 cm (1 in) wide and round or irregular; scales are peltate, brown, and woody when mature; seed cones mature in one growing season. Bark is gray to red-brown, fibrous, and stringy on small trees; the bark on large trees becomes more ridged with long, narrow plates. Large trees are more plated than baldcypress. The growth form is usually a smaller tree than baldcypress, but both species have a fluted and swollen trunk at the base and cypress knees.

HABITAT Flatwoods, pocosins, bays, and blackwater river swamps.

NOTES Although both cypress species can be found together, pondcypress is more common on nutrient-poor soils and tends to occur in stagnant waters. Other forest associates include Atlantic white-cedar,

From left to right:

Pondcypress leaves; photo by Lisa J. Samuelson.

Pondcypress seed cone.

water tupelo, and swamp tupelo. It is used for greenhouses, fencing, shingles, caskets, railroad ties, blinds, flooring, cabinetry, small boat construction, and boxes.

Taxodium is Latin for "yewlike," referring to the leaves; *ascendens* refers to the twigs. Pondcypress is one word because it is not a true cypress.

Baldcypress

Taxodium distichum (L.) Rich.

COMMON NAMES baldcypress, southern cypress

QUICK GUIDE Leaves flat, two-ranked, feathery, deciduous; cone round with shield-shaped scales; bark red-brown, fibrous, and stringy. Other distinguishing features include roots known as "cypress knees" growing out of the water and Spanish moss hanging from branches.

DESCRIPTION Leaves are deciduous, two-ranked, linear, about 1 cm (0.4 in) long, flat, soft, feathery, and green to yellow-green, turning red-brown in the autumn. Branchlets can be deciduous. The seed cone is round and up to 3.5 cm (1.4 in) wide; scales are peltate, brown and woody when mature; seed cones mature in one growing season. Bark is gray to red-brown, fibrous, and stringy; the bark of large trees has loose, narrow strips. The growth form is up to 36 m (120 ft) in height and 1.5 m (5 ft) in diameter. The base of the trunk is often swollen and fluted with woody roots extending out of the water like "knees" near the base.

HABITAT Swamps, ponds, river sloughs, and bottomlands.

NOTES Baldcypress can withstand prolonged flooding and can be found in pure stands, usually on soils with poor drainage, or with water tupelo, swamp tupelo, and pondcypress. This species is slow-growing and long-lived (up to 1,700 years), but few old-growth stands remain. The wood is yellow-white with reddish latewood, light, soft, decay resistant, and easily worked. It is used for greenhouses, fencing,

From left to right:

Baldcypress leaves.

Baldcypress seed cone.

shingles, caskets, railroad ties, blinds, flooring, cabinetry, small boat construction, and boxes. The seeds are eaten by waterfowl, such as wood ducks, and small mammals. The canopy is used as a nesting tree for birds such as osprey and bald eagles. Baldcypress is planted as an ornamental on a variety of sites.

Taxodium is Latin for "yew-like," referring to the leaves; *distichum* refers to the two-ranked leaves. Baldcypress is one word because it is not a true cypress.

Pine Family (*Pinaceae*)

Shortleaf Pine

Pinus echinata Mill.

COMMON NAMES shortleaf pine, southern yellow pine, short straw pine, rosemary pine

QUICK GUIDE Needles in bundles of two, short, flexible, not twisted; cone gray, with small sharp prickles; bark with red-brown plates, often with small resin holes.

DESCRIPTION Needles are in bundles of two or sometimes three, evergreen, 6–13 cm (2.4–5.1 in) long, green to yellow-green, and flexible. The seed cone is 4–7 cm (1.2–2.8 in) long, ovoid, flexible, and red-brown to gray, with small sharp prickles, and matures in two growing seasons. Bark is red-brown and rough on small trees; large trees have irregular orange-brown to red-brown plates, often with pin-sized resin holes in the plates. The growth form is up to 30 m (100 ft) in height and 1 m (3 ft) in diameter.

HABITAT Shortleaf pine has the widest range of all the southern pines and is found on a wide variety of upland sites.

NOTES Forest associates are numerous and include hickories, eastern redcedar, sweetgum, other pines, and a variety of oaks. Seedlings can sprout after damage from fire or grazing. Shortleaf pine is susceptible to littleleaf disease, which is caused by a fungus. The wood has

From left to right:

Shortleaf pine needles.

Shortleaf pine seed cones.

From left to right:

Shortleaf pine with resin holes in the bark.

Shortleaf pine bark.

Orange-gray scaly bark of sand pine for comparison

red-brown heartwood and yellow sapwood with prominent red-brown latewood, and is resinous and moderately heavy. Shortleaf pine is an important commercial species used for pulpwood, plywood, and construction lumber and was an important species for shipbuilding. The seeds are eaten by birds and small mammals, and the buds are eaten by small mammals such as the fox squirrel. The foliage provides excellent nesting, roosting, and loafing cover for a variety of bird species.

Pinus is Latin for "pine tree"; *echinata* means "prickly," referring to the cone.

SIMILAR SPECIES **Sand pine** (*Pinus clausa* [Chapm. ex Engelm.] Vasey ex Sarg.) is similar with needles in bundles of two that are short (5–9 cm [2.0–3.5 in] long), moderately twisted, and flexible; a cone with sharp prickles; and bark that is orange-gray with scaly plates. The cones may be embedded in the branches. It is distinguished from other two-needled pines by its limited range, habitat, and scrubby form. Sand pine is found in the coastal region of Alabama on sand dunes and on sandy, acidic, infertile soils with bluejack oak, turkey oak, sand post oak, slash pine, and longleaf pine. The Ocala (var. *clausa*) geographic race is found in northeast to south Florida and has serotinous cones. The Choctawhatchee (var. *immuginata*) geographic race is found in northwest Florida and extreme southwest Alabama, and has open cones at maturity. Sand pine is planted for pulpwood production, in Christmas tree plantations, and as a sand binder.

Slash Pine

Pinus elliottii Engelm.

COMMON NAMES slash pine, swamp pine, southern yellow pine

QUICK GUIDE Needles fascicled in bundles of two or three, long, flexible, moderately stout; cone shiny, chocolate brown, with weak prickles; bark with red-brown plates often with white or tan flecking.

DESCRIPTION Needles are in bundles of two or three, evergreen, 15–28 cm (5.9–11.0 in) long, shiny, moderately stout, and flexible. Twigs are stout, but not as stout as longleaf pine. The seed cone is 7–18 cm (2.8–7.1 in) long, ovoid, chocolate brown, and shiny, with weak prickles, and matures in two growing seasons. Bark is brown and rough on small trees; large trees have red-brown plates often with white-tan flecking. The growth form is 30 m (100 ft) in height and 1 m (3 ft) in diameter.

HABITAT Sandy soils with high moisture content, such as in depressions and bays, and on the edges of ponds and swamps, as well as on drier sites.

NOTES In natural stands, slash pine can be found growing with Atlantic white-cedar, loblolly-bay, water and swamp tupelo, red maple, sweetbay magnolia, longleaf pine, pond pine, and loblolly pine. Slash

From left to right:

Slash pine needles and seed cones; photo by Lisa J. Samuelson.

Slash pine seed cones.

From left to right:

Slash pine bark with pale flecking.

Slash pine bark with brown flecking.

pine is an important commercial species and is widely planted on a variety of sites in the lower Southern Coastal Plain of Alabama. The wood is comprised of red-brown heartwood and yellow sapwood with prominent red-brown latewood and is resinous and moderately heavy. It is grown in plantations to produce pulpwood, plywood, poles, and construction lumber. Slash pine was once a major source of turpentine and resin. The needles were used to make baskets, pine wool, and rough textiles. The seeds are eaten by birds and small mammals, and the buds are eaten by small mammals such as the fox squirrel. The foliage provides excellent nesting, roosting, and loafing cover for a variety of bird species.

Pinus is Latin for "pine tree"; *elliottii* is for Stephen Elliott, a botanist from South Carolina.

Spruce Pine

Pinus glabra Walter

COMMON NAMES spruce pine, cedar pine, Walter's pine

QUICK GUIDE Needles in bundles of two, short, slender, blue-green, flexible, sometimes twisted; cone with weak, deciduous prickles; bark furrowed rather than plated on large trees.

DESCRIPTION Needles are in bundles of two, evergreen, 5–10 cm (2.0–3.9 in) long, blue-green, thin, flexible, sometimes twisted, and fragrant (with an orange smell) when crushed. The seed cone is 4–8 cm (1.6–3.1 in) long, ovoid, and red-brown to gray, with weak and deciduous prickles, and matures in two growing seasons. Bark on small trees is gray to brown and smooth or shallowly ridged; on large trees the bark is brown to dark gray and furrowed. The growth form is up to 37 m (121 ft) in height and 1 m (3 ft) in diameter.

HABITAT Flatwoods, stream banks, bottomlands, hummocks, and on moist, fertile soils of upland sites.

NOTES Spruce pine is a shade-tolerant species found in mixed stands growing with red maple, sweetgum, American holly, sweet-bay magnolia, swamp tupelo, blackgum, slash pine, loblolly pine, laurel oak,

From left to right:

Spruce pine needles.

Spruce pine seed cones.

From left to right:

Bark on spruce pine is smooth or shallowly ridged.

Bark on large spruce pine is furrowed.

overcup oak, swamp chestnut oak, live oak, redbay, and American elm. The wood is light, soft, and brittle, and is sometimes used for lumber and pulpwood. The seeds are eaten by birds and small mammals, the buds are eaten by small mammals such as the fox squirrel, and the bark is eaten by beaver. The foliage provides excellent nesting, roosting, and loafing cover for a variety of bird species.

Pinus is Latin for "pine tree"; *glabra* means "lacking hair or smooth."

Longleaf Pine

Pinus palustris Mill.

COMMON NAMES longleaf pine, heart pine, southern yellow pine, long straw pine

QUICK GUIDE Needles in bundles of three, very long, moderately stout, flexible, in tufts at the ends of stout branches; terminal bud silver-white, stout; cone up to 25 cm (9.8 in) long; seedling grass stage.

DESCRIPTION Needles are in bundles of three, evergreen, 20–46 cm (7.9–18.1 in) long, dark green, shiny, moderately stout, flexible, and densely clustered in tufts on the ends of stout branches. The seed cone is cylindrical, 15–25 cm (5.9–9.8 in) long, and red-brown to gray, with slender and moderately sharp prickles, and matures in two growing seasons. Branches may curve upwards at the ends. Bark is gray-brown to orange-gray and rough on small trees; bark on large trees has orange-brown or red-brown plates that may become flaky. The growth form is up to 37 m (121 ft) in height and 1 m (3 ft) in diameter. Old trees become flat topped.

HABITAT A variety of sites, including flatwoods, sandhills, and swamp edges. Montane longleaf pine is found growing on mountain ridges in northern Alabama.

NOTES Longleaf pine is more fire tolerant than the other southern pines, and fire is important in maintaining longleaf pine ecosystems.

From left to right:

Longleaf pine needles; photo by Lisa J. Samuelson.

Longleaf pine seed cone.

Counterclockwise from upper left:

Longleaf pine pollen cones.

Longleaf pine grass-stage seedling.

Longleaf pine bark.

In general, longleaf pine can be longer lived (450 years) than the other southern pines (200–250 years). This is the only southern pine with large (up to 6 cm [2.4 in] long) silver-white buds. Longleaf pine has a grass stage, which can continue for many years, in which stem elongation is suppressed. When subjected to frequent fire, longleaf pine can be a dominant species, but without regular fire, hardwoods and other pine species become codominant or dominant. The wood has red-brown heartwood (heart pine) and yellow sapwood with prominent red-brown latewood and is resinous and moderately heavy. Wood uses include pulpwood, plywood, construction lumber, and utility poles. Longleaf pine was a main source of turpentine and resin. Wood from the stump and tap root are a source of "fatwood." The large, decay-resistant needles are highly desired for pine straw. The seeds are eaten by birds and small mammals, and some wildlife prefer its larger seed to that of other pines. The buds are eaten by small mammals such as the fox squirrel. The foliage provides excellent nesting, roosting, and loafing cover for a variety of bird species. Mature longleaf pine trees serve as nesting trees for the red-cockaded woodpecker.

Pinus is Latin for "pine tree"; *palustris* means "swamp."

Pond Pine

Pinus serotina Michx.

COMMON NAMES pond pine, bay pine, marsh pine, pocosin pine

QUICK GUIDE Needles in bundles of three, long, flexible; cone nearly round and serotinous with weak prickles; bark may show tufts of needles.

DESCRIPTION Needles are in bundles of three or sometimes four, evergreen, 13–21 cm (5.1–8.3 in) long, and flexible. The seed cone is ovoid when closed to nearly globose when open, 5–7 cm (2.0–2.8 in) long, serotinous (opening with fire or heat), and milk chocolate brown to gray, with weak prickles or lacking prickles, and matures in two growing seasons. The cone base may be embedded in the branch. Bark is gray-brown and rough on small trees; large trees have irregular red-brown and gray plates. Tufts of needles or sprouts may be seen sticking out of the bark. The growth form is up to 21 m (70 ft) in height and 61 cm (24 in) in diameter.

HABITAT Wet sites, swamps, ponds, flatwoods, pocosins, and bays.

NOTES Pond pine grows where fire is frequent, and after a fire sprouts can be seen emerging from the stem and branches. Pond pine can be found in pure stands or in mixed stands with red maple, Atlantic white-cedar, loblolly-bay, sweetgum, sweetbay magnolia, water and swamp tupelo, slash pine, loblolly pine, longleaf pine, live oak,

From left to right:

Pond pine needles.

Pond pine serotinous seed cones.

baldcypress, and pondcypress. This species is also considered a variety of pitch pine (*Pinus rigida* Mill.). The wood has red-brown heartwood and yellow sapwood with prominent red-brown latewood and is resinous and moderately heavy. It is used for pulpwood and construction lumber. The seeds are eaten by birds and small mammals, and the buds are eaten by small mammals such as the fox squirrel. The foliage provides excellent nesting, roosting, and loafing cover for a variety of bird species.

Pinus is Latin for "pine tree"; *serotina* means "late," referring to the serotinous cones.

Eastern White Pine

Pinus strobus L.

COMMON NAMES eastern white pine, white pine, northern white pine

QUICK GUIDE Needles in bundles of five, blue-green; cone long, slim, lacking prickles; bark red-brown, furrowed.

DESCRIPTION Needles are in bundles of five, evergreen, 8–14 cm (3.1–5.5 in) long, blue-green, slender, and flexible, and appear ethereal from a distance. The seed cone is oblong and slim, 8–20 cm (3.1–7.9 in) long, brown, and often curved in shape, with scales lacking prickles and often dotted with white resin; seed cones mature in two growing seasons. The branches are produced annually in whorls on the trunk. Bark is green-gray and smooth on small trees; large trees are gray-brown and furrowed with rough, narrow plates. The growth form is up to 46 m (150 ft) in height and 1 m (3 ft) in diameter.

HABITAT Moist, cool forests and near streams and rivers.

NOTES Eastern white pine is the only native eastern conifer with five needles per fascicle. It is one of the tallest trees in the eastern United States and long-lived (500 years). The USDA Plants Database currently lists eastern white pine in Lawrence, Winston, Tuscaloosa, Perry, and Lee Counties of Alabama, but there are reports of its occurrence in

From left to right:

Eastern white pine needles; photo by Lisa J. Samuelson.

Eastern white pine seed cones.

From left to right:

Smooth bark of young eastern white pine.

Furrowed bark of large eastern white pine.

other counties as well. Forest associates include red maple, American beech, Carolina silverbell, tulip-poplar, cucumbertree, black cherry, white oak, northern red oak, basswood, and eastern hemlock. Eastern white pine was one of the most important commercial species in the eastern United State before extensive logging. Virgin stands were once common in the eastern United States, but most were logged by the late 1800s. The wood is cream colored, soft, and light, and has a wide variety of uses including boxes, crates, millwork, cabinets, flooring, furniture, and matches. Eastern white pine is planted as an ornamental but requires moist, cool sites. It is a popular species for Christmas trees. The seeds are eaten by birds and small mammals, and the buds are eaten by small mammals such as the fox squirrel. The foliage provides excellent nesting, roosting, and loafing cover for a variety of bird species.

Pinus is Latin for "pine tree"; *strobus* refers to a fragrant, gum-yielding tree.

Loblolly Pine

Pinus taeda L.

COMMON NAMES loblolly pine, old field pine, meadow pine

QUICK GUIDE Needles in bundles of three, long, flexible; cone brown-gray with sharp prickles; bark with red-brown plates.

DESCRIPTION Needles are in bundles of three or sometimes four, evergreen, 12–23 cm (4.7–9.1 in) long, green to dark green, and flexible. The seed cone is conical, 6–14 cm (2.4–5.5 in) long, and brown-gray, with sharp prickles, and matures in two growing seasons. Bark is gray-black to red-brown and scaly on small trees; large trees have red-brown plates. The growth form is up to 30 m (100 ft) tall and 1 m (3 ft) in diameter.

HABITAT A variety of sites, including ridges, swamp edges, upland forests, old fields, and bottomlands.

NOTES Loblolly pine can form pure stands or be found in mixed stands with a wide variety of species. The wood has red-brown heartwood and yellow sapwood with prominent red-brown latewood and is resinous and moderately heavy. This species is the most important commercial species in the South and is planted for pulpwood,

From left to right:

Loblolly pine needles with the pollen and seed cones.

Loblolly pine seed cones.

plywood, and construction lumber. The seeds are eaten by birds and small mammals, and the buds are eaten by small mammals such as the fox squirrel. The foliage provides excellent nesting, roosting, and loafing cover for a variety of bird species. Mature trees are used as a nesting tree for the red-cockaded woodpecker.

Pinus is Latin for "pine tree"; *taeda* means "resinous," referring to the wood.

Virginia Pine

Pinus virginiana Mill. Carrière

COMMON NAMES Virginia pine, scrub pine

QUICK GUIDE Needles in bundles of two, short, twisted, stiff; cone scales with a thickened lip and slender, sharp prickles; bark orange-brown or red-brown and scaly.

DESCRIPTION Needles are in bundles of two, evergreen, 3–8 cm (1.2–3.1 in) long, stiff, twisted, and fragrant when crushed. The seed cone is 4–7 cm (1.6–2.8 in) long, ovoid, and red-brown to gray; scales bear a thick lip and a slender, sharp prickle; seed cones mature in two growing seasons. Bark is gray-brown to orange-brown or red-brown and scaly on small trees and becomes thin with red-brown or orange-brown plates on large trees. The growth form is usually only up to 23 m (75 ft) in height.

HABITAT A variety of upland sites, common on dry soils.

NOTES Virginia pine can form pure stands in old fields. In mixed stands, forest associates include red maple, upland hickories, eastern redcedar, sweetgum, blackgum, shortleaf pine, loblolly pine, and numerous upland oaks. The wood consists of red-brown heartwood and yellow sapwood with prominent red-brown latewood and is resinous.

From left to right:

Virginia pine needles with the pollen and seed cones.

Virginia pine seed cones.

From left to right:

Bark of young Virginia pine.

Bark of large Virginia pine.

Wood uses include pulpwood, construction lumber, and fuel. Virginia pine is planted in Christmas tree plantations and for privacy screens. The seeds are eaten by birds and small mammals, and the buds are eaten by small mammals such as the fox squirrel. The foliage provides excellent nesting, roosting, and loafing cover for a variety of bird species.

Pinus is Latin for "pine tree"; *virginiana* refers to the geographic range.

Eastern Hemlock

Tsuga canadensis (L.) Carrière

COMMON NAMES eastern hemlock, northern hemlock, Canadian hemlock

QUICK GUIDE Needles small, flat, in two rows, pegged at the base, with double white bands on the underside; cone small, up to 2 cm (0.8 in) long, lacking prickles; bark dark red-brown and deeply furrowed.

DESCRIPTION Needles are evergreen, flat, 8–17 mm (0.3–0.7 in) long, somewhat two-ranked, and dark shiny green or yellow-green; apex is rounded or notched; base is pegged; underside has two narrow white bands. The seed cone is 1.3–2.0 cm (0.5–0.8 in) long, ovoid to round, brown, and born on the ends of twigs; scales lack prickles; seed cones mature in one growing season. Bark is dark cinnamon-brown and scaly or shallowly grooved on smaller trees; large trees become deeply furrowed with purple inner bark. The crown has pendulous, graceful branches. The growth form is a graceful tree up to 37 m (121 ft) in height and 1.2 m (4 ft) in diameter. Old-growth trees may be much larger.

HABITAT Moist to damp soils such as the edges of streams, in sheltered coves, and on northern slopes.

From left to right:

Eastern hemlock needles and seed cones.

Smaller cones of eastern hemlock (top two) compared to the larger cones of Carolina hemlock (bottom three).

From left to right:

Bark of young eastern hemlock.

Very thick bark of large eastern hemlock.

NOTES Eastern hemlock is more common in the northern part of the state. It is a very shade-tolerant and long-lived tree (up to 900 years). Forest associates include red maple, sugar maple, white ash, tulip-poplar, black birch, Carolina silverbell, cucumbertree, eastern white pine, black cherry, white oak, northern red oak, and basswood. The wood is pale red-brown, light, and moderately hard, and is used for pulpwood, containers, and construction lumber. The bark was once a major source of tannin. Eastern hemlock is planted as an ornamental, but it is heat intolerant and susceptible to insect damage. Its dense canopy provides important cover for white-tailed deer, birds, and small mammals. The seeds are eaten by birds and small mammals, and the foliage is browsed by white-tailed deer, squirrel, and rabbit. The hemlock wooly adelgid inadvertently introduced from Asia is causing widespread mortality and the future of this species is of great concern.

Tsuga is Japanese for "hemlock"; *canadensis* refers to the geographic range.

SIMILAR SPECIES Carolina hemlock. (*Tsuga carolinina* Engelm.) is similar and found in the southern Appalachian mountains as far south as Georgia, usually on drier sites than eastern hemlock, and has longer needles (1–2 cm [0.4–0.8 in] long) arranged more spirally around the twig rather than two-ranked and a larger seed cone (2–4 cm [0.8–1.6 in] in length, see the seed cones photograph for comparison).

Moscatel Family (*Adoxaceae*)

Rusty Blackhaw

Viburnum rufidulum Raf.

COMMON NAMES rusty blackhaw, blue haw, rusty nannyberry

QUICK GUIDE Leaves opposite, simple, oval to obovate, margin finely serrate, underside rusty pubescent; twigs with rusty pubescence; fruit a blue drupe in red-stalked clusters; bark blocky.

DESCRIPTION Leaves are opposite, simple, deciduous, oval to obovate, 5–10 cm (2.0–3.9 in) long, and shiny green; apex is rounded or notched; base is cuneate to rounded; margin is finely serrate; underside has rusty pubescence; autumn color is purple-red. The petiole is grooved and red, with rusty pubescence, and is often winged. Twigs are slender and green-brown with rusty pubescence when young but become red-brown to gray and glabrous when older; leaf scar is U-shaped with three bundle scars. The terminal bud is acute, about 1 cm (0.4 in) long; scales are valvate, rusty colored, scurfy, and pubescent. Flowers are perfect and white, and bloom in flat-topped clusters in early spring. Fruit is a drupe, nearly round, about 1 cm (0.4 in) wide, dark blue, and red stalked, and matures in drooping clusters in early summer. Bark is dark brown and blocky on large trees. The growth form is a shrub or small tree up to 6 m (20 ft) in height.

HABITAT Moist to dry sites, mostly in well-drained woodlands.

From left to right:

Rusty blackhaw leaves.

Rusty blackhaw fruit; courtesy of Nancy Loewenstein.

Rusty blackhaw leaves and flowers.

Clockwise from left:

Rusty blackhaw bark.

Margin of southern arrowwood leaves for comparison.

Smooth blackhaw leaves for comparison.

NOTES Rusty blackhaw is moderately browsed by white-tailed deer and beaver. The fruit is eaten by numerous songbirds and game birds, including northern bobwhite and wild turkey. The fruit is also eaten by coyote, white-tailed deer, beaver, skunk, and small mammals. The viburnums in general are attractive landscaping trees, and many cultivars are available.

Viburnum is the Latin name of wayfaring-tree of Eurasia; *rufidulum* (rufus) refers to the red-brown pubescence on leaves, twigs, and buds.

SIMILAR SPECIES Other species of viburnums that can be found in Alabama are small trees or shrubs. The list includes **witherod** (*Viburnum cassinoides* L.), which is found in the northern part of the state on moist to wet sites and is identified by elliptical leaves from 5–15 cm (2.0–5.9 in) long with an irregularly serrate or dentate-crenate margin and by flattened, orange, scurfy buds. **Southern arrowwood** (*Viburnum dentatum* L.) is found on a variety of sites and has nearly round leaves that are 5–10 cm (2.0–3.9 in) long with large dentate or serrate teeth on the margin. **Possumhaw** (*Viburnum nudum* L.) is found on mesic sites and has elliptical leaves with a mostly entire margin, a midrib and petiole with rusty pubescence, and twigs that are rusty scurfy. **Smooth blackhaw** (*Viburnum prunifolium* L.) is an occasional, small tree found on well-drained soils and has elliptical to oval leaves with a finely serrate margin, mostly glabrous leaf underside, and glaucous twigs.

Sweetgum Family (*Altingiaceae*)

Sweetgum

Liquidambar styraciflua L.

COMMON NAMES sweetgum, red gum, satin walnut

QUICK GUIDE Leaves alternate, simple, star shaped, with five to seven deeply palmate lobes, margin finely serrate; fruit a spiny ball of capsules; bark gray-brown with flattened interlacing ridges; branches often with corky wings.

DESCRIPTION Leaves are alternate, simple, deciduous, star shaped, 8–15 cm (3.1–5.9 in) wide, and palmately veined, with five to seven palmate lobes; apices are acuminate; base is truncate; margin is finely serrate; petiole is 13–15 cm (5.1–5.9 in) long; autumn color is yellow or red. Twigs are moderately stout, yellow-brown to green-brown, and shiny, with lenticels and often corky wings; leaf scar is shield shaped with three bundle scars. The terminal bud is up to 2 cm (0.8 in) long and acute; scales are overlapping, red-brown to purple-green, shiny, and tipped with white pubescence. Flowers are imperfect and appear with the leaves in the spring; staminate flowers are in yellow-green, round heads in erect terminal clusters up to 6 cm (2.4 in) long; pistillate flowers are in small green heads on long drooping stalks. Fruit is a spiny woody ball ("gumballs") of capsules about 3 cm (1.2 in) wide,

Clockwise from upper left:

Sweetgum leaf.

Sweetgum flowers.

Sweetgum fruit.

Sweetgum bark.

and matures in early fall. Bark is gray-brown with corky ridges on small trees and darker with flat interlacing ridges on large trees. The growth form is up to 36 m (120 ft) in height and 1.2 m (4 ft) in diameter, and can form thickets.

HABITAT A variety of sites ranging from dry sandy soils to intermittently flooded bottomlands.

NOTES Sweetgum is a vigorous colonizer and root sprouter. Forest associates are numerous and site dependent. The wood is moderately heavy and moderately hard and is used for veneer, trim, boxes, pallets, and pulpwood. The red heartwood is used for furniture. The sap was used in chewing gums and skin balms. Cultivars have been developed to enhance fall color and cold hardiness, and a fruitless cultivar is available. Seeds are used by mallard ducks and eaten directly from seed balls by various finches and occasionally by gray squirrels. The bark is relished by beavers.

Liquidambar refers to the brown-yellow sap; *styraciflua* refers to the fragrant, gummy resin (styrax) exuded from cut twigs.

Cashew Family (*Anacardiaceae*)

American Smoketree

Cotinus obovatus Raf.

COMMON NAMES American smoketree, chittamwood

QUICK GUIDE Leaves alternate, simple, obovate to oval, base and apex rounded, margin entire, petiole purple-red; fruit stalks create a smoky appearance in midsummer; bark gray-brown and scaly.

DESCRIPTION Leaves are alternate, simple, deciduous, obovate to oval, and 10–15 cm (3.9–5.9 in) long; apex is rounded or notched; base is rounded; margin is entire or repand; petiole is up to 5 cm (2 in) long and purple-red; leaf underside has green pubescence on veins; autumn color is bright red, orange, or yellow. Twigs are purple-gray to red-brown and glaucous, with lenticels; leaf scar is shield shaped to crescent shaped and bluish around the scar. The terminal bud is about 3 mm (0.1 in) long and acute; scales are overlapping, red-brown, and pubescent. Flowers are dioecious, small, and clustered on long stalks in panicles; petals are yellow-green; anthers are bright yellow; flowers bloom in the spring after the leaves emerge. Fruit is a drupe,

Clockwise from upper left:

American smoketree leaves.

American smoketree flowers.

American smoketree fruit.

American smoke-
tree bark.

compact and kidney shaped, on feathery/hairy pinkish stalks in mid-
to late summer. The stalks without fruit create a smoky appearance.
Bark is red-brown to gray-brown and thin, with loose scales, and is
shaggier on large trees. The growth form is a shrub or small tree up
to 10 m (35 ft) in height, and sometimes forms thickets.

HABITAT Uncommon, found on rocky slopes and limestone soils in
northern Alabama.

NOTES American smoketree wood is used for fence posts, and an orange dye was extracted from the wood during the Civil War. The Eu-
ropean species is often planted as an ornamental.

Cotinus is derived from Greek for "wild-olive"; *obovatus* refers to the
leaves.

Winged Sumac

Rhus copallinum L.

COMMON NAMES winged sumac, shining sumac, flameleaf sumac

QUICK GUIDE Leaves alternate and pinnately compound, with 9 to 21 leaflets; leaflets shiny green, margin mostly entire, rachis green winged; twigs velvety pubescent; fruit a conical collection of dull red drupes at branch ends in late summer.

DESCRIPTION Leaves are alternate, pinnately compound, deciduous, and up to 30 cm (11.8 in) long with 9 to 21 leaflets. Leaflets are lanceolate or falcate, 4–7 cm (1.6–2.4 in) long, and shiny green; margin is mostly entire; rachis is green winged; autumn color is bright red. Twigs are slender with brown pubescence and raised lenticels; leaf scar is horseshoe shaped with numerous bundle scars, and nearly encircles the lateral bud. A true terminal bud is lacking; lateral bud is round, about 3 mm (0.1 in) long, light brown, and fuzzy. Flowers are dioecious or polygamous, greenish-white, and clustered in dense panicles at twig ends in the summer. Fruit is a drupe, about 3 mm (0.1 in) wide, pubescent, dull red, and grows in dense conical clusters up to 20 cm (7.9 in) long at branch ends; fruit matures in the fall. Fruit clusters persist over the winter. Bark is gray-brown and smooth with lenticels, and becomes warty or scaly on large stems. The growth

From left to right:

Winged sumac leaf.

Winged sumac flowers.

Winged sumac fruit.

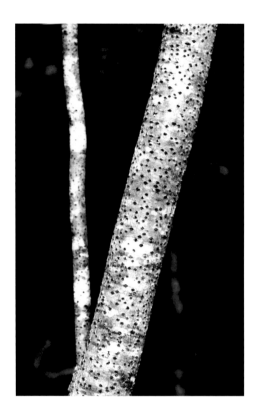

Winged sumac bark.

form is a shrub or small tree up to 6 m (20 ft) in height with an open crown and often forms thickets due to root sprouting.

HABITAT Common in old fields and disturbed areas and along wood edges and power lines.

NOTES Winged sumac is found in pure stands or thickets or in open areas with eastern redcedar, sweetgum, water oak, and other tree species that colonize open areas. The leaves and flowers contain tannin. Beaver, foxes, squirrels, and rabbits will eat the bark, particularly in winter. The fruit is eaten by many songbirds and game birds, including northern bobwhite, wild turkey, woodpeckers, thrushes, and vireos. The fruit is also eaten by white-tailed deer and opossum. The crushed fruit has a tart flavor and some soak it in water, strain it, and then mix with sugar to make a sort of lemonade. Cultivars have been developed to enhance form and fall color.

Rhus is from "rhous," the Greek name for sumac; *copallinum* is Spanish for "varnishlike sap."

Smooth Sumac

Rhus glabra L.

COMMON NAMES smooth sumac, scarlet sumac

QUICK GUIDE Leaves alternate, pinnately compound, with 11 to 31 leaflets and a purple-red, glabrous rachis; leaflets with a serrate margin; twigs glaucous; fruit a conical collection of bright red drupes at branch ends in the fall.

DESCRIPTION Leaves are alternate, pinnately compound, deciduous, and up to 40 cm (15.7 in) long, with 11 to 31 leaflets. Leaflets are lanceolate to falcate and 5–14 cm (2.0–5.5 in) long; margin is serrate; rachis is purple-red and glabrous; autumn color is bright red. Twigs are stout, glaucous, purple-red, and sticky when cut; leaf scar is heart shaped with numerous bundle scars, and nearly encircles the lateral bud. A true terminal bud is lacking; lateral bud is round, about 5 mm (0.2 in) long, light brown, and fuzzy. Flowers are dioecious or polygamous, greenish-white, and clustered in dense panicles at twig ends

Clockwise from upper left:

Smooth sumac form.

Smooth sumac leaf.

Smooth sumac fruit.

Smooth sumac flowers.

From left to right:

Smooth sumac bark.

Staghorn sumac velvety fruit for comparison.

in the summer. Fruit is a drupe, about 3 mm (0.1 in) wide, pubescent, bright red, and grows in dense conical clusters up to 20 cm (15.7 in) long at branch ends. Fruit clusters are usually larger in smooth sumac than in winged sumac. The fruit matures in the fall and persists over the winter. Bark is gray-brown and smooth with lenticels, and becomes warty and scaly on large stems. The growth form is a shrub or small tree up to 6 m (20 ft) in height with an open crown and often forms dense thickets due to root sprouting.

HABITAT Common in old fields and cutover areas and along wood edges.

NOTES Smooth sumac is found in pure stands or thickets or in open areas with eastern redcedar, sweetgum, water oak, and other tree species that colonize open areas. The leaves and flowers contain tannin. Beaver, foxes, squirrels, and rabbits will eat the bark, particularly in winter. The fruit is eaten by many songbirds and game birds, including northern bobwhite, wild turkey, woodpeckers, thrushes, and vireos. The fruit is also eaten by white-tailed deer and opossum. The crushed fruit has a tart flavor and some soak it in water, strain it, and then mix with sugar to make a sort of lemonade. Cultivars have been developed to enhance form and fall color.

Rhus is from "rhous," the Greek name for sumac; *glabra* means "hairless," referring to the hairless rachis and twig.

SIMILAR SPECIES **Staghorn sumac** (*Rhus typhina* L.) is sometimes found in the northern part of the state. It has a pubescent rachis and velvety hairy twigs, fruit, and buds, but the leaflets have a serrate margin similar to smooth sumac.

Poison-Sumac

Toxicodendron vernix (L.) Kuntze

COMMON NAMES poison-sumac, swamp-sumac, poison-elderberry, thunderwood

QUICK GUIDE Leaves alternate, pinnately compound, with 7 to 13 elliptical leaflets, margin entire, petiole and rachis bright red; fruit a whitish drupe in open, drooping clusters; bark gray and smooth, often with dark patches.

DESCRIPTION Leaves are alternate, pinnately compound, deciduous, and up to 40 cm (15.7 in) long, with 7 to 13 leaflets. Leaflets are elliptical to oblong and 5–10 cm (3.0-3.9 in) long; margin is entire; petiole and rachis are bright red; autumn color is bright red. Twigs are stout, brown-gray, and glabrous, with lenticels; leaf scar is large and shield or heart shaped, with prominent bundle scars. The terminal bud is about 1.5 cm (0.6 in) long and conical, with two purple-brown scales. Flowers are dioecious or polygamous and in arching panicles from leaf axils after the leaves; petals are yellow-green; anthers are orange. Fruit is a drupe, round, about 6 mm (0.2 in) wide, gray-white to white, juicy, and grows in open dangling clusters that persist over winter. Bark is mottled, brown-gray, and smooth, with lenticels or warts, and often showing black patches from exudation of sap. The growth form is a shrub or small tree up to 10 m (33 ft) in height with an open crown.

HABITAT Stream edges and swamps, bogs, pineland depressions.

From left to right:

Poison-sumac leaf.

Poison-sumac fruit.

Poison-sumac
bark.

NOTES Caution! The oleoresin in all plant parts of poison-sumac may cause a severe rash in humans when touched. The smoke is also dangerous when inhaled. The fruit is eaten by birds.

Toxicodendron means "poison tree"; *vernix* means "varnish" and refers to the clear sap, which turns black when exposed to air.

Custard-Apple Family (*Annonaceae*)

Pawpaw

Asimina triloba (L.) Dunal

COMMON NAMES common pawpaw, wild banana, Indian banana

QUICK GUIDE Leaves alternate, simple, large, obovate, with a green pepper smell when crushed, underside with a velvety maroon pubescence; flower petals in threes and purple-maroon; fruit banana-like.

DESCRIPTION Leaves are alternate, simple, deciduous, obovate to oblong, 10–30 cm (3.9–11.8 in) long, and often drooping, with prominent venation and a green pepper smell when crushed; apex and base are acute; margin is entire; autumn color is yellow. Twigs are slender and light brown, with soft maroon pubescence; leaf scar is raised and crescent shaped, with numerous bundle scars. The buds are naked and covered with velvety maroon pubescence; terminal bud is conical and about 1 cm (0.4 in) long; flower buds are plump. Flowers are perfect, petals are in groups of three, with two layers of three purple-maroon petals with prominent venation; flower is up to 5 cm (2 in) wide when fully open and blooms before or with the leaves. Fruit is a large berry, up to 13 cm (5.1 in) long, thick, and often curved; fruit contains large, disc-shaped seeds; fruit matures in the summer. Bark is brown-gray, mottled, and smooth with warts. The growth form is a shrub or understory tree up to 12 m (40 ft) in height.

From left to right:

Pawpaw leaves.

Pawpaw flowers.

Pawpaw fruit.

HABITAT Moist, fertile soils near streams and in bottomland forests and occasionally on drier sites.

NOTES Pawpaw is usually an understory tree or shrub, typically growing in clusters. The inner bark was used for string and fishnets, and the fruit has a tropical flavor and was used in custards. The flesh is edible but the seeds are not. Some people may be allergic to the fruit, however. The fruit is relished by raccoon, opossum, foxes, squirrels, birds, and small rodents. Larvae of the zebra swallowtail butterfly feed on the foliage. Pawpaw is an interesting landscape tree due to flowers and fruit, but it will root sprout.

Asimina is believed to be a Native American word meaning "sleeve-shaped fruit"; *triloba* refers to the petals in groups of three.

SIMILAR SPECIES **Dwarf pawpaw** (*Asimina parviflora* [Michx.] Dunal) is found on mesic and xeric sites throughout the state. It has smaller leaves (up to 20 cm [7.9 in] long), flowers (2 cm [0.8 in] wide when open), and fruit (up to 8 cm [3.1 in] long), and leaves have an acuminate apex.

Holly Family (*Aquifoliaceae*)

Large Gallberry

Ilex coriacea (Pursh) Chap.

COMMON NAMES large gallberry, sweet gallberry

QUICK GUIDE Leaves alternate, simple, evergreen, obovate or oval, margin thick with an occasional bristle-tipped tooth above the middle; fruit a black drupe; bark gray with warty lenticels.

DESCRIPTION Leaves are alternate, simple, evergreen, elliptical to obovate or oval, 4–9 cm (1.6–3.5 in) long, dark green, and shiny; apex is bristle tipped; margin is thickened and entire or with an occasional bristle-tipped tooth mostly above the middle; underside is glabrous or pubescent. Twigs are gray-green to dark brown, pubescent, and sometimes sticky; leaf scar is semicircular with one bundle scar. The buds are small; scales are green or red-brown, overlapping, and pubescent. Flowers are dioecious, small, and green-white, and bloom in late spring. Fruit is a drupe, round, 7 mm (0.3 in) wide, shiny black, and sweet, and contains several ribbed stones. The fruit matures in the autumn and usually does not persist over winter. Bark is brown-gray to green and smooth, with warty lenticels. The growth form is a shrub or small tree to up to 6 m (20 ft) in height.

From left to right:

Large gallberry leaves with spiny teeth above the middle.

Large gallberry fruit.

From left to right:

Large gallberry bark.

Margin of ink-berry leaves for comparison.

HABITAT Sandy and wet soils of swamps, bogs, bays, bottoms, and seeps.

NOTES The fruit of large gallberry is eaten by birds and small mammals, and the leaves are browsed by deer. Bees visit the flowers.

Ilex refers to *Quercus ilex*, or holly oak, an evergreen Mediterranean tree with toothed leaves; *coriacea* means "leathery," referring to the leaves.

SIMILAR SPECIES **Inkberry** (*Ilex glabra* [L.] A. Gray) is a shrub found on similar sites and is distinguished by leaves with an entire margin that are sometimes serrate rather than spiny above the middle and by a black, dry, bitter drupe that persists over winter.

Possumhaw

Ilex decidua Walter

COMMON NAMES possumhaw, deciduous holly

QUICK GUIDE Leaves alternate, simple, deciduous, elliptical, margin serrate to crenate, clustered on spur branches; fruit a bright red, juicy drupe prominent after leaf fall; bark smooth and gray with warts.

DESCRIPTION Leaves are alternate, simple, deciduous, elliptical to obovate or oblanceolate, 4–8 cm (1.6–3.1 in) long, and clustered on spur branches; apex is acute to rounded; base is cuneate to rounded; margin is serrate or crenate; midrib is pubescent; autumn color is yellow. Twigs are green-brown to gray-white and glabrous and have lenticels; leaf scar is semicircular with one bundle scar. The buds are small; scales are overlapping, green to brown or red-brown, and pubescent or glabrous. Flowers are dioecious, small, and green-white, and bloom in late spring. Fruit is a drupe, round, 6–10 mm (0.2–0.4 in) wide, orange-red to bright red, and juicy, and contains several ribbed stones. The fruit matures in the autumn and may persist until the following spring. Bark is thin, green-gray to gray-brown, mottled, and mostly smooth with warts but may be rougher on larger individuals. The growth form is a shrub or small tree up to 6 m (20 ft) in height but sometimes taller.

From left to right:

Possumhaw leaves and flowers.

Possumhaw fruit.

Possumhaw bark.

HABITAT Swamps and floodplains, on moist upland soils, occasionally on drier sites.

NOTES Possumhaw is a browse for white-tailed deer, and the fruit is eaten by various songbirds and game birds, including northern bobwhite and wild turkey. The fruit is also eaten by white-tailed deer and gray squirrel.

Ilex refers to *Quercus ilex*, or holly oak, an evergreen Mediterranean tree with toothed leaves; *decidua* refers to the deciduous leaves.

SIMILAR SPECIES Other deciduous hollies include **Sarvis holly** (*Ilex amelanchier* Curtis ex Chapm.), which is rare and found in the extreme southern part of the state; it is identified by deciduous obovate to oblanceolate leaves with a serrate margin above the middle and a densely pubescent underside and by a red drupe. **Carolina holly** (*Ilex ambigua* [Michx.] Torr.), also known as sand holly, is found on well-drained sandy soils of the Southern Coastal Plain and identified by deciduous leaves less than 10 cm (3.9 in) long with an acute apex, a leaf margin entire or minutely serrate or crenate above the middle, and a shiny red drupe 5–10 mm (0.2–0.4 in) wide.

American Holly

Ilex opaca Aiton

COMMON NAMES American holly, white holly

QUICK GUIDE Leaves alternate, simple, evergreen, margin with spiny and widely spaced teeth; fruit a bright red drupe in the autumn; bark smooth, thin, and gray mottled.

DESCRIPTION Leaves are alternate, simple, evergreen (persisting three years), elliptical to obovate, 5–12 cm (2.0–4.7 in) long, dark green, leathery, and thick; margin has sharp, spiny, and widely spaced teeth, or is occasionally entire. Twigs are gray-white to brown, mottled, pubescent when young, and have lenticels; leaf scar is semicircular with one bundle scar. The buds are small; scales are overlapping, green or red, and pubescent. Flowers are dioecious, small, and green-white, and bloom in late spring. Fruit is a drupe, round, 4–10 mm (0.2–0.4 cm) wide, and red or orange-red, and contains several ribbed stones. The fruit matures in the autumn and persists over the winter. Bark is light gray, mottled, and smooth, with warts. The growth form is up to 15 m (50 ft) in height.

HABITAT A variety of sites, including swamps, bottomlands, hammocks, stream edges, well-drained uplands, and occasionally on drier, protected sites. This species can tolerate salt spray.

NOTES American holly is usually an understory tree found with red maple, American beech, white and green ash, sweetgum, tulip-poplar, blackgum, white oak, and water oak. The wood is ivory, hard, heavy, and close-grained, and it is used for veneer, turnery, specialty items

From left to right:

American holly leaves and fruit.

American holly staminate flowers.

American holly pistillate flowers.

American holly bark.

(including piano keys and scientific instruments), furniture, and cabinets. The fruit is eaten by northern bobwhite, wild turkey, numerous songbirds (including cedar waxwings), and small to midsize mammals. Branches with foliage and fruit are popular Christmas decorations. Though slow growing, American holly is a popular ornamental for large spaces for the evergreen foliage and red fruit.

Ilex refers to *Quercus ilex*, or holly oak, an evergreen Mediterranean tree with toothed leaves; *opaca* means "dark or shaded," referring to the dark green, thick leaves.

Yaupon

Ilex vomitoria Aiton

COMMON NAME yaupon, black drink holly

QUICK GUIDE Leaves alternate, simple, evergreen, oblong to oval, small, margin crenate; fruit a bright orange-red drupe; bark smooth and white-gray.

DESCRIPTION Leaves are alternate, simple, evergreen, oblong to oval, 1–4 cm (0.4–1.6 in) long, and shiny; margin crenate. Twigs are red-brown and pubescent; older twigs are gray-white, stiff, thornlike, and often at 90-degree angles; leaf scar is semicircular with one bundle scar. Buds are small; scales are overlapping, green or red, and pubescent. Flowers are dioecious, small, and green-white, and bloom in late spring. Fruit is a drupe, round, 4–8 mm (0.2–0.3 in) wide, and orange-red, and contains several ribbed stones. The fruit matures in the autumn. Bark is thin, white-gray to gray-brown, mottled, and smooth. The growth form is a shrub or small tree up to 6 m (20 ft) in height.

HABITAT Dry or wet sandy sites, including flatwoods, upland forests, and coastal forests.

NOTES Yaupon is planted as an ornamental for its evergreen foliage and bright fruit, but it does root sprout. The fruit is eaten by various songbirds and small mammals, and the foliage is browsed by white-tailed deer.

Ilex refers to *Quercus ilex*, or holly oak, an evergreen Mediterranean tree with toothed leaves; *vomitoria* refers to the reported ability

From left to right:

Yaupon leaves and flowers.

Yaupon leaves and fruit.

of the leaves to induce regurgitation when consumed. Some Native American tribes used a drink derived from this tree in purification ceremonies.

SIMILAR SPECIES Other evergreen hollies that may be confused with yaupon include **myrtle-leaved holly** (*Ilex myrtifolia* Walter), which is found in the Southern Coastal Plain of Alabama and identified by shiny, lanceolate to oblanceolate, small (0.5–3.0 cm [0.2–1.2 in] long), and sessile leaves with a mostly entire margin or an occasional bristle-tipped tooth, and obscure veins, and by a red drupe. **Dahoon** (*Ilex cassine* L.) is found in the Southern Coastal Plain of Alabama in swamps and on stream edges and has leaves that are 4–10 cm (1.6–3.9 in) long and elliptical in shape with a rigid spine at the apex and margin with an occasional small tooth. The twigs are downy pubescent and the fruit is orange-red.

Ginseng Family (*Araliaceae*)

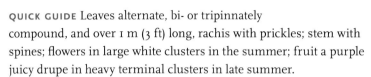

Devil's Walkingstick

Aralia spinosa L.

COMMON NAMES devil's walkingstick, Hercules-club, angelica-tree, prickly-ash

QUICK GUIDE Leaves alternate, bi- or tripinnately compound, and over 1 m (3 ft) long, rachis with prickles; stem with spines; flowers in large white clusters in the summer; fruit a purple juicy drupe in heavy terminal clusters in late summer.

DESCRIPTION Leaves are alternate, bi- or tripinnately compound, deciduous, and up to 1.6 m (5.2 ft) long, and form a large diamond-shaped silhouette; petiole and rachis are stout and purple-red; rachis has prickles. Leaflets are mostly ovate and 5–10 cm (2.0–3.9 in) long; apex is acute to acuminate; base is rounded or inequilateral; margin is serrate; midrib is prickly; autumn color is purple-red. Twigs are stout, gray-brown, and have spines and lenticels; leaf scar is U-shaped and lined with spines. The terminal bud is up to 2 cm (0.8 in) long and conical or ovoid; scales are overlapping, brown-gray, and rough edged. Flowers are perfect or imperfect, with green-white petals, and clustered at stem and branch ends in the summer. Fruit is a drupe, round, 5–8 mm (0.2–0.3 in) wide, purple-black, and juicy, and

Clockwise from upper left:

Devil's walking-stick leaf.

Devil's walking-stick flowers.

Devil's walking-stick fruit.

From left to right:

Devil's walking-stick young stem.

Bark of large devil's walkingstick.

born on purple-red stalks in drooping terminal clusters that mature in late summer and early fall. Bark is gray-brown, with spines, and on larger trees it is more furrowed. The growth form is a single-stemmed shrub or small tree up to 10 m (33 ft) in height, often forming thickets due to root sprouting.

HABITAT A variety of sites, including upland forests and edges of swamps and streams.

NOTES The fruit of devil's walkingstick is eaten by numerous songbirds (a favorite of cedar waxwings) and small to medium-size mammals such as eastern chipmunk and striped skunk, and the foliage is moderately browsed by white-tailed deer. Bees and other insects visit the flowers. The flowers and fruit make this an interesting natural border plant, but it can be invasive.

Aralia is from a Native American name; *spinosa* means "bearing spines."

Birch Family (*Betulaceae*)

Hazel Alder

Alnus serrulata (Aiton) Willd.

COMMON NAMES hazel alder, common alder, tag alder, brookside alder

QUICK GUIDE Leaves alternate, simple, obovate, margins finely toothed, underside with maroon or pale pubescence; buds stalked and with maroon valvate scales; fruit a nutlet in a woody cone; bark brown-gray, sinewy, and smooth.

DESCRIPTION Leaves are alternate, simple, deciduous, obovate to elliptical, and 5–13 cm (2.0–5.1 in) long; apex is obtuse; base is rounded to cuneate; margin is finely serrate and somewhat wavy; lateral veins are sunken and end in a margin tooth; underside has maroon or pale pubescence; autumn color is yellow or brown. Twigs are slender, brown to gray, and pubescent; leaf scar is half-round with three bundle scars. A true terminal bud is lacking; lateral bud is stalked, about 8 mm (0.3 in) long, and has two or three scales that are red-maroon and valvate. Flowers are imperfect and in separate catkins; mature staminate catkins are red-yellow, about 8 cm (2.7 in) long, and dangle; pistillate catkins are erect, short, and red when they open in the spring. Fruit is a nutlet, laterally winged, and enclosed in a woody cone about 1 cm (0.4

From left to right:

Hazel alder leaves and fruit.

Hazel alder flowers.

Clockwise from left:

Hazel alder bark.

Wavy margin and lopsided base of witch-hazel leaves for comparison.

Unique stringy yellow petals of witch-hazel flowers for comparison.

in) long. The fruit matures in the fall, and the cones persist over the winter. Bark is brown-gray, mottled, smooth, fluted, and sinewy. The growth form is a shrub or small tree up to 6 m (20 ft) in height and forms dense thickets on water margins.

HABITAT Stream and pond edges, and in drainage ditches, swamps, and wet woods.

NOTES Hazel alder thickets can easily take over pond and stream edges so its landscape value is limited. The foliage is a browse for beaver, rabbits, muskrat, and white-tailed deer, and the fruit is eaten by songbirds. Beavers use the stems to build dams. Dense thickets are valuable as wildlife cover.

Alnus is Latin for "alder"; *serrulata* refers to the finely serrate leaf margin.

SIMILAR SPECIES **Witch-hazel** (*Hamamelis virginiana* L.) is a shrub or small tree found throughout Alabama on moist soils, and its leaves are sometimes confused with hazel alder. Witch-hazel leaves have a wavy margin and inequilateral base, and are often lopsided in shape; the terminal bud is stalked, flattened, and naked; the flowers have four stringy yellow or reddish petals and bloom in the autumn; and the fruit is a woody urn-shaped capsule.

Black Birch

Betula lenta L.

COMMON NAMES black birch, sweet birch, cherry birch

QUICK GUIDE Leaves alternate, simple, ovate, two-ranked, base cordate, margin serrate or doubly serrate; twigs glabrous with a strong wintergreen smell when cut; bark black and smooth on young trees, with scaly plates on large trees.

DESCRIPTION Leaves are alternate, simple, deciduous, two-ranked, ovate 5–13 cm (2.0–5.1 in) long, and underside with pubescent veins; apex is acute; base is cordate or inequilateral; margin is serrate or doubly serrate; autumn color is yellow. Twigs zigzag and are slender, flexible, maroon, glabrous, and shiny, with lenticels and a strong wintergreen odor and taste when cut. The leaf scar is crescent shaped with three bundle scars. A true terminal bud is lacking; lateral bud is acute, about 6 mm (0.2 in) long, and divergent; scales are green-brown to yellow-brown, mostly glabrous, and shiny. Flowers are imperfect; staminate catkins are long and drooping; pistillate catkins are shorter, upright, and thick. Fruit is a nutlet that is laterally winged

Black birch leaves and fruit.

From left to right:

Bark of a young black birch.

Bark of a large black birch.

Golden bark of a young yellow birch for comparison.

and in an upright papery cone up to 4 cm (1.6 in) long; fruit scale is three-lobed with mostly glabrous bracts; fruit matures in early fall. Bark is maroon to black and smooth with horizontal lenticels on young trees; large trees have gray to black scaly plates. The growth form is up to 18 m (60 ft) in height and 61 cm (2 ft) in diameter.

HABITAT Moist, cool forests.

NOTES Black birch is found in north Alabama growing with yellow buckeye, red maple, sugar maple, tulip-poplar, eastern white pine, black cherry, American elm, eastern hemlock, white oak, and northern red oak. The wood is blond to red-brown, straight-grained, heavy, and hard, and is used for veneer, furniture, paneling, turnery, cabinets, and specialty items. Black birch is a source of wintergreen oil and birch beer. The foliage is a browse for white-tailed deer, and a variety of songbirds and small mammals eat the fruit.

Betula is the Latin name for "birch"; *lenta* means "pliant," referring to the flexible twigs.

SIMILAR SPECIES Yellow birch (*Betula alleghaniensis* Britton) is similar and occurs in Tennessee counties bordering northern Alabama. It is distinguished by shiny golden to bronze peeling bark on younger trees, a weak wintergreen odor and taste from cut twigs, pubescent bracts on the fruit scale, and lateral buds appressed to the twig.

River Birch

Betula nigra L.

COMMON NAMES river birch, red birch, water birch

QUICK GUIDE Leaves alternate, simple, two-ranked, triangular, base truncate, margin doubly serrate; bark with silver, red-brown or black scales, may peel revealing pale pink or orange inner bark.

DESCRIPTION Leaves are alternate, simple, deciduous, two-ranked, triangular or ovate, and 4–10 cm (1.6–3.9 in) long; apex is acuminate; base is wedge shaped; margin is doubly serrate; petiole is pubescent or glabrous; underside is glabrous or pubescent; autumn color is yellow. Twigs zigzag and are slender, maroon, and pubescent or glabrous, with lenticels. The leaf scar is triangular with three bundle scars. A true terminal bud is lacking; lateral bud is conical to triangular, about 6 mm (0.2 in) long, and appressed; scales are yellow-maroon, pubescent or glabrous, and overlapping. Flowers are imperfect; staminate catkins are yellow-green, long, and drooping; pistillate catkins are upright and thick. Fruit is a nutlet, laterally winged, in an upright papery cone up to 4 cm (1.6 in) long; fruit scale has three lobes; fruit matures in late spring or early summer. Bark is scaly with silver, red-brown, or gray-brown irregular scales. On young trees or the upper stem of large trees, the bark may peel into wide strips exposing pale

From left to right:

River birch leaves.

River birch flowers.

River birch fruit.

From left to right:

Peeling bark of river birch.

Scaly bark of river birch.

pink or orange inner bark. The growth form is up to 24 m (80 ft) in height and 1 m (3 ft) in diameter.

HABITAT Edges of rivers and streams, often leaning out over the water, and in bottoms.

NOTES Forest associates of river birch include boxelder, red maple, silver maple, sugarberry, hackberry, ashes, sycamore, eastern cottonwood, bottomland oaks, black willow, and American elm. The wood is blond to red-brown, straight-grained, moderately heavy, and moderately hard, and is used for pulpwood, baskets, and inexpensive furniture. The foliage is a browse for white-tailed deer, and a variety of songbirds and small mammals eat the fruit. River birch is planted to control erosion and as an ornamental for the attractive bark and form. Cultivars have been developed to enhance the bark colors and disease resistance.

Betula is the Latin name for "birch"; *nigra* means "dark," referring to the bark.

Hornbeam

Carpinus caroliniana Walter

COMMON NAMES hornbeam, American hornbeam, blue beech, musclewood, ironwood

QUICK GUIDE Leaves alternate, simple, two-ranked, thin, margin doubly serrate, lateral veins usually unbranched; buds maroon and white striped; fruit a nutlet attached to a three-lobed leafy bract; bark gray, smooth, sinewy, and fluted.

DESCRIPTION Leaves are alternate, simple, deciduous, two-ranked, ovate to elliptical, 5–10 cm (2.0–3.9 in) long, and thin; apex is acute to acuminate; base is cordate; margin is doubly serrate; lateral veins are usually unbranched; petiole and underside are pubescent; autumn color is a dull orange-red to brown. Twigs zigzag and are slender, maroon, and pubescent or glabrous, with white lenticels. The leaf scar is crescent shaped to round with three bundle scars. A true terminal

Clockwise from upper left:

Hornbeam leaves.

Hornbeam pistillate flowers.

Hornbeam staminate flowers.

Hornbeam fruit.

Hornbeam bark.

bud is lacking; lateral bud is ovoid and about 4 mm (0.2 in) long; scales are maroon and white striped and overlapping. Flowers are perfect and appear with the leaves; staminate catkins are up to 4 cm (1.6 in) long and drooping; pistillate flowers have forked red stigmas and are obscure at the tips of new shoots. Fruit is a nutlet in a three-lobed leafy bract that hangs in pairs and matures in late summer. Bark is thin, gray or blue-gray, smooth, fluted, and sinewy. The growth form is a small understory tree usually only reaching 9 m (30 ft) in height.

HABITAT Near streams, in floodplains and swamps, and on cool slopes in many forest cover types.

NOTES Hornbeam is a small tree usually found in the understory. The wood is whitish, hard, and dense, and is used for specialty items such as tool handles and bowls. The foliage is a browse for white-tailed deer, rabbit, and beaver, and the seed is eaten by songbirds, wild turkey, squirrels, and rodents.

Carpinus is Latin for "hornbeam"; *caroliniana* refers to the geographic range.

Hophornbeam

Ostrya virginiana (Mill.) K. Koch

COMMON NAMES hophornbeam, eastern hophornbeam, ironwood

QUICK GUIDE Leaves alternate, simple, two-ranked, thin, margin doubly serrate, lateral veins divide at the leaf margin; buds green and brown striped; fruit hoplike; bark red-brown and shreddy.

DESCRIPTION Leaves are alternate, simple, deciduous, two-ranked, ovate to lanceolate, 5–13 cm (2.0–5.1 in) long, and thin; apex is acute or acuminate; base is cordate; margin is doubly serrate; lateral veins divide near the leaf margin; underside and petiole are pubescent; autumn color is yellow. Twigs zigzag and are red-brown, slender, and pubescent or glabrous; leaf scar is crescent shaped to round with three bundle scars. A true terminal bud is lacking; lateral bud is ovoid-acute, about 6 mm (0.2 in) long, and divergent; scales are overlapping and green and brown striped, or sometimes red-brown. Flowers are imperfect and bloom in the spring; staminate catkins are up to 7 cm (2.8 in) long and drooping; pistillate catkins have obscure forked red stigmas at the tips of new shoots. Fruit is a nutlet enclosed in a hoplike inflated sac that is about 2 cm (0.8 in) long, and matures in late summer. Bark is gray-brown to red-brown and scaly or shreddy. The

From left to right:

Hophornbeam leaves.

Hophornbeam staminate and pistillate flowers.

Hophornbeam fruit.

Hophornbeam bark.

growth form is usually only up to 12 m (40 ft) in height but can be larger in bottoms.

HABITAT A wide variety of sites and soils but common near streams, in floodplains, and on moist cool slopes in many forest cover types.

NOTES Hophornbeam is usually an understory tree found with a variety of forest associates. The wood is whitish and hard and is used for mallets, tool handles, specialty items, and fence posts. The seed is eaten by wild turkey, songbirds, and small to midsize mammals. It is occasionally browsed by white-tailed deer. The buds and catkins are eaten in winter by many birds, squirrels, and rabbits.

Ostrya is Latin for "a tree with hard wood"; *virginiana* refers to the geographic range.

Catalpa Family (*Bignoniaceae*)

Southern Catalpa

Catalpa bignonioides Walter

COMMON NAMES southern catalpa, catawba, cigar tree, Indian-bean, fish-bait tree

QUICK GUIDE Leaves whorled, simple, large, heart shaped, margin entire, underside pubescent; twigs stout with moon crater–like leaf scars; flowers white with maroon and yellow spots; fruit a long, beanlike capsule; bark brown-gray and scaly.

DESCRIPTION Leaves are whorled or opposite, simple, deciduous, and heart shaped, with a 13–20 cm (5.1–7.9 in) long blade and long (about 13 cm [5.1 in]) petiole; apex is acute or abruptly acuminate; base is cordate or rounded; margin is entire or sometimes lobed at the base; underside is pubescent; autumn color is yellow. Twigs are stout, light brown, and glabrous or lightly pubescent, with lenticels; leaf scar is round and raised, resembles moon craters, and varies in size at a given node; bundle scars are in a circular pattern. A true terminal bud is lacking; lateral bud is moundlike and embedded and has light brown scales. Flowers are perfect and trumpetlike, with white petals showing maroon and yellow spots. Flowers bloom in large erect panicles in early summer. Fruit is a capsule, usually up to 30 cm (11.8 in) long but sometimes longer, green, two-valved, and beanlike; seeds have pointed and fringed wings; fruit matures in early fall and persists over the winter. Bark is gray-brown and scaly. The growth form is up to 15 m (50 ft) in height with a low open crown.

From left to right:

Southern catalpa leaf.

Southern catalpa flowers.

Southern catalpa fruit.

HABITAT Rich soils of riverbanks, floodplains, and swamp edges.

From left to right:

Southern catalpa bark.

Northern catalpa flowers.

NOTES Southern catalpa is found growing with boxelder, river birch, water hickory, sycamore, water oak, willow oak, and American elm. It can be found growing in open areas and near old homesteads. The wood is grayish, soft, durable, and coarse-grained, and is used for railroad ties, fence posts, and specialty items. It was used for telegraph poles. The catalpa sphinx larva (catawba worms) that feeds on the leaves is used for fish bait. Cultivars are planted as ornamental trees for the attractive leaves and flowers and tolerance to stress, but the fruit can be messy.

Catalpa is from a Cherokee Indian name; *bignonioides* refers to the flower, which is similar to the trumpet-flower in the genus *Bignonia*.

SIMILAR SPECIES Northern catalpa (*Catalpa speciosa* Warder ex Engelm.) has naturalized throughout the eastern United States and is identified by bark that is more ridged, flowers with fewer purple spots, and a larger fruit (up to 46 cm [18.1 in] long).

Strawberry-Shrub Family (*Calycanthaceae*)

Sweetshrub

Calycanthus floridus L.

COMMON NAMES sweetshrub, Carolina allspice, strawberry-bush, hairy sweetshrub

QUICK GUIDE Leaves opposite, simple, ovate to elliptical, margin entire, prominent venation, underside white, with a spicy odor when crushed; flowers red-maroon with curved petals, strawberry scented; fruit in a leathery pod.

DESCRIPTION Leaves are opposite, simple, deciduous, two-ranked, ovate to elliptical, 5–13 cm (2.0–5.1 in) long, and dark green, and have a spicy odor when crushed; apex is acute to acuminate; base is acute; margin is entire; petiole is up to 1.5 cm (0.6 in) long; upper surface has prominent venation; underside is white and pubescent; autumn color is yellow. Twigs are brown to red-brown, slender, flattened at the node, and fragrant when cut, with lenticels. The leaf scar is U-shaped with three bundle scars. A true terminal bud is lacking; lateral bud is small, ovoid, naked, and hairy. Flowers are dark red to maroon, 2.5–4.0 cm (1.0–1.6 in) across, and sweet-smelling, and have many curved petals and sepals. The flowers bloom from spring to early summer. Fruit is an achene, about 6 mm (0.2 in) wide, chocolate colored, and hairy. Mutiple achenes are contained in a long-stalked, fibrous, wrinkled, and drooping pod that is up to 8 cm (3.1 in) long and persists

From left to right: Sweetshrub leaves. Sweetshrub leaf underside

Clockwise from upper left:

Sweetshrub flower.

Sweetshrub fruit.

Sweetshrub seedling; photo by Lisa J. Samuelson.

Sweetshrub young stem; photo by Lisa J. Samuelson.

over winter. Bark is brown and thin, has lenticels, and is fragrant when cut. The growth form is usually a shrub but may grow up to 4 m (13 ft) tall and grows in clusters due to root sprouting.

HABITAT A variety of sites, usually woodlands.

NOTES Sweetshrub leaves are browsed by deer but other wildlife uses are limited. The fragrant flowers were used to scent cabinets and drawers. Sweetshrub is a popular ornamental shrub and cultivars with yellow flowers are available.

Calycanthus refers to the similar sepals and petals; *floridus* means flowering.

Cannabis and Hop Family (*Cannabaceae*)

Sugarberry

Celtis laevigata Willd.

COMMON NAMES sugarberry, southern hackberry, sugar hackberry, Mississippi hackberry

QUICK GUIDE Leaves alternate, simple, ovate to lanceolate, with three main veins, apex acuminate, margin entire or irregularly serrate; twigs and buds pubescent; fruit an orange-red to purple drupe; bark gray-brown with corky warts or smooth.

DESCRIPTION Leaves are alternate, simple, deciduous, ovate to lanceolate, and 5–13 cm (2.0–5.1 in) long, and have three main veins that arise from the base; apex is acuminate; base is rounded and inequilateral; margin is entire or irregularly serrate above the middle; upper side is smooth but sometimes scabrous; autumn color is yellow. Twigs zigzag and are slender and red-brown to green, with fine pubescence and lenticels; leaf scar is crescent shaped with three bundle scars. A true terminal bud is lacking; lateral bud is triangular, about 3 mm (0.1 in) long, and appressed; scales are overlapping, red-black, and pubescent. Flowers are imperfect and perfect, small, and yellow-green or green-white, and lack petals; flowers bloom in the spring with the leaves. Fruit is a drupe, round, about 7 mm (0.3 in) wide, orange-red to purplish, and sweet, and matures in the fall. Bark is gray-brown to blue-gray with corky or woody warts; the bark of large trees is smooth. The growth form is usually only up to 15 m (50 ft) in height but can be as tall as 30 m (100 ft) on good sites.

From left to right:

Sugarberry leaves.

Sugarberry flowers.

Sugarberry fruit.

Sugarberry bark; photo by Lisa J. Samuelson.

HABITAT Occasionally in uplands and on limestone soils but more common in bottomlands and edges of streams and swamps.

NOTES Forest associates include bitternut hickory, hackberry, ash, sweetgum, eastern cottonwood, bottomland oaks, black willow, and American elm. The wood is yellow-gray, moderately hard, and moderately heavy, and is used for furniture, veneer, and boxes. The entire fruit is edible and mildly sweet, but the seeds are chalky. The fruit is a favorite of many birds, including the American robin, yellow-bellied sapsucker, mockingbird, mourning dove, and quail. The fruit is also used by a variety of small to midsize mammals.

Celtis was a name for African lotus, which has sweet fruit; *laevigata* means "smooth," perhaps referring to the leaves or bark.

Hackberry

Celtis occidentalis L.

COMMON NAMES hackberry, northern hackberry, common hackberry

QUICK GUIDE Leaves alternate, simple, ovate, with three main veins, apex acute or acuminate, base cordate, margin irregularly serrate; twigs and buds mostly glabrous; fruit a red-purple drupe; bark gray-brown with warty, corky ridges.

DESCRIPTION Leaves are alternate, simple, deciduous, ovate, and 5–13 cm (2.0–5.1 in) long, and with three main veins arising from the base; apex is acute to acuminate; base is cordate and inequilateral; margin is serrate; upper side is smooth or scabrous; autumn color is yellow. Twigs zigzag and are slender, green-brown to red-brown, and mostly glabrous, with lenticels. The leaf scar is crescent shaped with three bundle scars. A true terminal bud is lacking; lateral bud is triangular, about 6 mm (0.2 in) long, and appressed; scales are overlapping, red-black, and mostly glabrous. Flowers are perfect and imperfect, small, and yellow-green or green-white, lack petals, and bloom in the spring with the leaves. Fruit is a drupe, round, about 8 mm (0.3 in) wide, and red-purple, and matures in the fall. Bark is gray-brown to ash gray with warty corky ridges; large trees have scaly ridges. The growth form is up to 15 m (50 ft) in height.

HABITAT Mostly in bottomlands but also on limestone soils, slopes, and bluffs and in uplands.

From left to right:
Hackberry leaves.
Hackberry flowers.
Hackberry fruit.

Hackberry bark.

NOTES Forest associates of hackberry include sugar maple, green ash, sweetgum, American beech, sycamore, and American elm. The wood is yellow-gray, moderately hard, and moderately heavy, and is used for furniture, veneer, and boxes. The entire fruit is edible and mildly sweet, but the seeds are chalky. The fruit is a favorite of many birds, including the American robin, yellow-bellied sapsucker, mockingbird, mourning dove, and quail. The fruit is also used by a variety of small to midsize mammals.

Celtis was a name for African lotus, which has sweet fruit; *occidentalis* refers to the Western Hemisphere.

Georgia Hackberry

Celtis tenuifolia Nutt.

COMMON NAMES Georgia hackberry, dwarf hackberry

QUICK GUIDE Leaves alternate, simple, ovate, with three main veins, apex acute or acuminate, base cordate, margin irregularly serrate above the middle, upper surface very scabrous; fruit an orange-red drupe; bark gray-brown with corky warts.

DESCRIPTION Leaves are alternate, simple, deciduous, ovate, and 4–7 cm (1.6–2.8 in) long; apex is acute or acuminate; base is cordate and inequilateral, with three main veins arising from it; margin is irregularly serrate above the middle or entire; upper side is very scabrous; autumn color is yellow. Twigs zigzag and are slender, green-brown, and mostly glabrous, with lenticels. The leaf scar is crescent shaped with three bundle scars. A true terminal bud is lacking; lateral bud is triangular, about 3 mm (0.1 in) long, and appressed; scales are red-black, overlapping, and glabrous or pubescent. Flowers are perfect and imperfect, small, and yellow-green or green-white, lack petals, and bloom in the spring with the leaves. Fruit is a drupe, round, about 7 mm (0.3 in) wide, and orange-red to red-brown, and matures in the fall. Bark is gray-brown with corky warts. The growth form is a shrub or small tree up to 9 m (30 ft) in height.

HABITAT Understory of upland forests and well-drained soils.

From left to right:

Georgia hackberry leaves and immature fruit.

Georgia hackberry flowers.

NOTES The wood of Georgia hackberry is not commercially important. The entire fruit is edible and mildly sweet, but the seeds are chalky. The fruit is a favorite of many birds, including the American robin, yellow-bellied sapsucker, mockingbird, mourning dove, and quail. The fruit is also used by a variety of small to midsize mammals.

Celtis was a name for African lotus, which has sweet fruit; *tenuifolia* means "thin leaved."

Dogwood Family (*Cornaceae*)

Alternate-Leaf Dogwood

Cornus alternifolia L. f.

COMMON NAME alternate-leaf dogwood

QUICK GUIDE Leaves alternate but may appear opposite, simple, ovate, with arcuate venation; flowers in white flat-topped clusters; fruit a purple-black drupe in red-stalked clusters; bark green and white streaked on young stems.

DESCRIPTION Leaves are alternate (but clustered at twig ends so may appear opposite or whorled), simple, deciduous, ovate to oval or elliptical, and 5–14 cm (2.0–5.5 in) long; apex is acute to acuminate; base is rounded to cuneate; margin is minutely serrate or entire and wavy; venation is arcuate; petiole is long and often red; underside is pubescent; autumn color is yellow, red, or maroon. Twigs are slender, bright green to maroon, and glabrous; leaf scar is V-shaped with three bundle scars. The terminal bud is about 6 mm (0.2 in) long, and ovoid, with two to three glabrous or finely pubescent scales. Flowers are perfect, white, and compact, and bloom in flat-topped clusters after the leaves in late spring. Fruit is a drupe, round, 5–10 mm (0.2–0.4 in) wide, purple-black, and grows in erect red-stalked clusters in the

Clockwise from left:

Alternate-leaf dogwood leaves.

Alternate-leaf dogwood flowers.

Alternate-leaf dogwood fruit.

Alternate-leaf
dogwood bark.

summer. Bark is green-brown, smooth, and vertically streaked on small trees, and gray-brown and ridged on large trees. The growth form is a shrub or small tree up to 9 m (30 ft) in height.

HABITAT Rich, moist soils of upland forests and edges of streams and swamps.

NOTES The foliage of alternate-leaf dogwood is browsed by white-tailed deer, and the fruit is eaten by birds and small mammals.

Cornus is Latin for "horn," referring to the hard wood; *alternifolia* refers to the alternate leaves.

Flowering Dogwood

Cornus florida L.

COMMON NAMES flowering dogwood, flowering cornel, common dogwood

QUICK GUIDE Leaves opposite, simple, ovate, with arcuate venation; twigs bright green or maroon; flowers with four white petal-like bracts in early spring; fruit a bright red drupe that appears in the fall; bark dark and shallowly blocky.

DESCRIPTION Leaves are opposite, simple, deciduous, mainly ovate or oval, and 7–13 cm (2.8–5.1 in) long, with arcuate venation and fibrous hairs when split; apex is acute; base is rounded; margin is entire; underside is pubescent; autumn color is yellow, red, or maroon. Twigs are slender, bright green or maroon, and glabrous; new growth "telescopic" to previous growth; leaf scar is V-shaped with three bundle scars. The terminal vegetative bud is acute, and about 6 mm (0.2 in) long, with two green valvate and pubescent scales; the flower bud is larger, pink-green, and onion shaped. Flowers are perfect and grow in yellow-green clusters surrounded by four white, petal-like, notched bracts; flowers bloom before and with the leaves. Fruit is a drupe, ovoid, about 1 cm (0.4 in) long, shiny red, and grows in erect clusters. Fruit matures in late summer or early fall. Bark is gray-black and shallowly blocky. The growth form is up to 12 m (40 ft) in height and 30 cm (1 ft) in diameter.

HABITAT Stream edges and moist, upland forests.

NOTES Flowering dogwood is an understory tree found growing with a wide variety of species. The wood is reddish-brown, hard, and heavy,

From left to right:

Flowering dogwood leaves.

Flowering dogwood flower.

Flowering dogwood fruit.

Flowering dog-
wood bark.

and is used for tool handles, mallet heads, and golf club heads, and in woodcarving. A red dye and a quininelike drug were derived from the bark. The foliage is browsed by white-tailed deer, and beaver eat the bark. The fruit is eaten by a large number of birds and mammals, including waterfowl, northern bobwhite, wild turkey, numerous song-birds, foxes, squirrels, and black bear. Because of its wide distribution and palatability, the fruit is a staple food for wildlife. Flowering dog-wood is a popular ornamental but suffers from insects and disease if not planted on moist, acidic soils and in partial shade. Cultivars offer a range of flower colors.

Cornus is Latin for "horn," referring to the hard wood; *florida* means "abounding in flowers."

SIMILAR SPECIES Other *Cornus* species that can be found in Alabama are commonly multistem shrubs or small trees. **Swamp dogwood** (*Cornus foemina* Mill.) is found in wet areas and is similar to flowering dog-wood, but the flowers are in white flat-topped clusters, fruit is a purple-blue drupe in long-stalked clusters, and bark is ridged rather than blocky. Leaves of swamp dogwood are glabrous or with pale pubescence on the underside. **Silky dogwood** (*Cornus amomum* Mill.) is also found on wet sites and is similar to swamp dogwood but the leaf underside and twig have pale or rusty pubescence and the twig has a brown rather than white pith. **Eastern roughleaf dogwood** (*Cornus asperifolia* Michx.) is very rare and identified by scabrous leaves with erect hairs on the underside, flowers in compact flat-topped clusters, and white fruit.

Water Tupelo

Nyssa aquatica L.

COMMON NAMES water tupelo, cotton-gum, water-gum, tupelo gum

QUICK GUIDE Leaves alternate, simple, elliptical, large, margin entire or with an occasional large tooth, petiole long and pubescent; fruit a large purple-black drupe with a prominently ridged stone; bark gray-brown with scaly or blocky ridges; trunk swollen at the base.

DESCRIPTION Leaves are alternate, simple, deciduous, elliptical to oblong or ovate to obovate, and the blade is 7–30 cm (2.8–11.8 in) long; apex is acute to acuminate; base is rounded to cuneate; margin is entire or with an occasional large dentate tooth; petiole is long (3–5 cm [1.2–2.0 in]), grooved, and pubescent; underside is pale and pubescent; autumn color is yellow. It is called cotton-gum because of the cotton-like hair on unfolding leaves. Twigs are stout, red-brown, glabrous or pubescent, and diaphragmed, with lenticels; leaf scar is heart shaped with three bundle scars. The terminal bud is ovoid to round, and about 3 mm (0.1 in) long, with green or red overlapping scales. Flowers are imperfect or perfect and yellow-green, and bloom in the spring with the leaves; staminate flowers are clustered in round heads; pistillate flowers have a prominent style and stigma born alone on long (4 cm [1.6 in]) stalks. Fruit is a drupe, ovoid, about 3 cm (1.2 in) long, purple-black, white speckled, and juicy, on long, drooping stalks; the stone has about 10 longitudinal ridges; fruit matures in late

From left to right:

Water tupelo leaves.

Shorter petioles of Ogeechee tupelo leaves and its fruit for comparison.

Water tupelo fruit.

From left to right:

Water tupelo bark.

Water tupelo form and habitat.

summer to early fall. Bark is gray-brown and shallowly grooved with scaly, blocky, or flattened ridges. The trunk is swollen at the base. The growth form is up to 30 m (100 ft) in height and 1 m (3 ft) in diameter.

HABITAT Deep swamps, wet flats, and floodplain forests.

NOTES Water tupelo is found in pure stands or with red maple, water hickory, swamp cottonwood, waterlocust, swamp tupelo, overcup oak, baldcypress, and pondcypress. The wood is white to brown-gray, cross-grained, moderately heavy, and moderately hard, and is used for pulpwood, pallets, crates, boxes, and furniture. The flowers are popular with honey bees, and the nectar is a source of tupelo honey. The fruit is eaten by a wide variety of songbirds and game birds, including northern bobwhite, wild turkey, red-cockaded woodpecker, and wood ducks. Black bear, squirrels, foxes, opossum, and raccoon also eat the fruit and the bark is eaten by beaver.

Nyssa refers to a water-loving nymph in classical mythology; *aquatica* refers to the wet habitat.

SIMILAR SPECIES Ogeechee tupelo (*Nyssa ogeche* Bartram ex Marshall) has been reported in one county in southeast Alabama. It is similar to water tupelo, but the petiole is shorter, the leaf is more rounded than that of water tupelo, and the stone of the drupe has papery longitudinal wings. Honey made from the Ogeechee tupelo is highly prized.

Swamp Tupelo

Nyssa biflora Walter

COMMON NAMES swamp tupelo, swamp blackgum, two-flowered tupelo

QUICK GUIDE Leaves alternate, simple, many elliptical, margin entire, smaller than water tupelo; fruit a drupe, purple-black, with a ridged stone, often in pairs; bark gray-brown and shallowly grooved.

DESCRIPTION Leaves are alternate, simple, deciduous, elliptical to lanceolate or oblanceolate, and 4–15 cm (1.6–5.9 in) long; apex is acute; base is mostly cuneate; margin is entire; underside is glabrous or has fine pubescence; autumn color is bright red. Twigs are slender, red-brown, glabrous, and diaphragmed, with lenticels; leaf scar is almost round with three bundle scars. The terminal bud is acute and about 4 mm (0.2 in) long, with green or red overlapping scales. Flowers are imperfect or perfect, and yellow-green, and bloom in the spring after the leaves; staminate flowers are clustered in round heads; pistillate flowers have a prominent style and stigma and usually appear in pairs on long stalks. Fruit is a drupe, nearly round, 1.0–1.5 cm (0.4–0.6 in) long, blue-black, juicy, and usually in pairs on long stalks; the stone is prominently ribbed; fruit matures in late summer to early fall. Bark is gray-brown and shallowly grooved and becomes somewhat blocky or has flattened ridges on large trees. The growth form is up to 24 m (80 ft) in height and 1 m (3 ft) in diameter with a swollen base.

From left to right:

Swamp tupelo leaves.

Swamp tupelo fruit in pairs.

From left to right:

Swamp tupelo
base and bark.

Swamp tupelo
bark.

HABITAT Frequently inundated bottomlands and swamps.

NOTES Swamp tupelo is found with red maple, Atlantic white-cedar, loblolly-bay, sweetbay magnolia, water tupelo, pond pine, baldcypress, and pondcypress. The wood is white to brown-gray, cross-grained, moderately heavy, and moderately hard, and is used for pulpwood, pallets, crates, boxes, and furniture. The flowers are popular with honey bees and the nectar is a source of tupelo honey. The fruit is eaten by a wide variety of songbirds and game birds, including northern bobwhite, wild turkey, red-cockaded woodpecker, and wood ducks. Black bear, squirrels, foxes, opossum, and raccoon also eat the fruit and the bark is eaten by beaver.

Nyssa refers to a water-loving nymph in classical mythology; *biflora* refers to the flowers being born in pairs.

Blackgum

Nyssa sylvatica Marshall

COMMON NAMES blackgum, black tupelo, tupelo-gum, sourgum

QUICK GUIDE Leaves alternate, simple, many obovate, margin entire but saplings sometimes with a few dentate teeth near the apex; leaf scar with three bundle scars; fruit a drupe, blue-black, with a shallowly ridged stone; bark variable, commonly gray-brown and thickly ridged or blocky.

DESCRIPTION Leaves are alternate, simple, deciduous, elliptical to obovate or oval, and 5–16 cm (2.0–6.3 in) long; apex is acute or abruptly acuminate; base is cuneate to blunt; margin is entire or has three to five dentate teeth near the apex (mostly on saplings); underside is glabrous or has a fine pubescence; autumn color is bright red. Twigs are slender, red-brown, glabrous, and diaphragmed, with lenticels; leaf scar is almost round with three bundle scars. The terminal bud is ovoid and about 6 mm (0.2 in) long; scales are overlapping, red-brown to green-brown, and glabrous or have pale or golden pubescence. Flowers are imperfect or perfect and yellow-green and bloom

Clockwise from upper left:

Blackgum leaves.

Blackgum staminate flowers.

Blackgum fruit.

Blackgum pistillate flowers.

From left to right:

Blackgum blocky bark.

Blackgum bark with flattened ridges.

in the spring after the leaves; staminate flowers are clustered in round heads; pistillate flowers have a prominent style and stigma and are arranged in groups of up to five at the end of a long stalk. Fruit is a drupe, ovoid, about 1.3 cm (0.5 in) long, blue-black, and juicy, with a shallowly ribbed stone. There are usually three to five drupes on a long stalk, and the fruit matures in late summer to early fall. Large branches are often at a 90-degree angle to the trunk. Bark is gray-brown and highly variable, ranging from deeply grooved and blocky to shallowly grooved with flattened or scaly ridges.

The growth form is up to 24 m (80 ft) in height and 1 m (3 ft) in diameter

HABITAT A wide variety of sites ranging from dry uplands to edges of streams and swamps.

NOTES Blackgum is a component of many forest cover types and common throughout Alabama. The wood is white to gray-brown, moderately hard, and moderately heavy, and is used for pulpwood, veneer, containers, pallets, railroad ties, woodenware, and gunstocks. The fruit is eaten by a wide variety of songbirds and game birds, including northern bobwhite, wild turkey, red-cockaded woodpecker, and wood ducks. Black bear, squirrels, foxes, opossum, and raccoon also eat the fruit, and the bark is eaten by beaver. The flowers are visited by honey bees and the nectar is a source of tupelo honey. Of all the *Nyssa* species, blackgum is the most important to wildlife because of its widespread distribution. It is an attractive ornamental tree because of its good form, brilliant fall color, and tolerance of a range of conditions.

Nyssa refers to a water-loving nymph in classical mythology; *sylvatica* means "of the forest," referring to the ubiquitous range of this species.

Ebony Family (*Ebenaceae*)

Common Persimmon

Diospyros virginiana L.

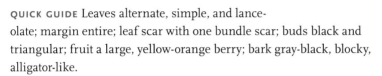

COMMON NAMES common persimmon, possum-wood, simmon, American persimmon

QUICK GUIDE Leaves alternate, simple, and lance-olate; margin entire; leaf scar with one bundle scar; buds black and triangular; fruit a large, yellow-orange berry; bark gray-black, blocky, alligator-like.

DESCRIPTION Leaves are alternate, simple, deciduous, lanceolate to elliptical or oblong, and 5–15 cm (2.0–5.9 in) long, often with black spots in the autumn; apex is acute or abruptly acuminate; base is rounded; margin is entire; underside is mostly glabrous; autumn color is yellow. Twigs zigzag and are brown and glabrous or pubescent, with orange lenticels; leaf scars are elevated and crescent shaped, with one slitlike bundle scar; old bud scales often remain at branch junctions. A true terminal bud is lacking; lateral bud is black, triangular (like a "snake head"), and about 3 mm (0.1 in) long, with two overlapping scales. Flowers are mostly dioecious, yellow-green, and appear after the leaves in late spring; pistillate flowers are bell shaped and solitary from leaf axils; staminate flowers are urn shaped and appear in clusters of two to four. Fruit is a berry, round, about 4 cm (1.6 in) wide, pulpy, and yellow to orange, with watermelon-like seeds. The

From left to right:

Common persimmon leaves and flowers.

Common persimmon fruit.

From left to right:

Common persimmon bark.

Common persimmon bark with orange in the grooves.

fruit is edible and tastes like apricot when mature but has an astringent, unpleasant taste when immature. Bark is shallowly grooved with orange in the grooves on small trees; large trees are gray-black, thick, and very blocky. The growth form is up to 24 m (80 ft) in height and 60 cm (23.6 in) in diameter.

HABITAT A variety of sites ranging from bottomlands to dry sandy soils.

NOTES Forest associates of common persimmon are numerous and site dependent. The wood is white to light brown, heavy, and hard, and is used for furniture, veneer, and turnery. The fruit is relished by raccoon, foxes, white-tailed deer, opossum, coyote, skunks, and the domestic dog (my dogs love it!). The large seeds of the fruit are eaten by wild turkey, northern bobwhite, and small rodents. The flowers are visited by bees. The fruit is used in jams and wine. Common persimmon has limited value in landscaping due to leaf spot, but cultivars with seedless fruit and brilliant fall foliage have been developed.

Diospyros means "divine fruit"; *virginiana* refers to the geographic range.

Heath Family (*Ericaceae*)

Sourwood

Oxydendrum arboreum (L.) DC.

COMMON NAMES sourwood, sorrel tree,
lily-of-the-valley tree

QUICK GUIDE Leaves alternate, simple, elliptical,
sour tasting, margin with fine teeth, midrib with hairs; fruit a stiff,
drooping cluster of capsules persistent over winter; bark dark and
deeply grooved with orange-red within the grooves.

DESCRIPTION Leaves are alternate, simple, deciduous, mostly elliptical, 10–18 cm (3.9–7.1 in) long, and sour tasting; apex and base are
acute; margin has fine teeth; midrib has hairs; autumn color is bright
red. Twigs zigzag and are yellow-green or red and mottled, with black
lenticels; leaf scar is elevated and shield shaped, with one bundle scar.
A true terminal bud is lacking; lateral bud is brown, small (about 1
mm [0.1 in] long), round, and embedded. Flowers are perfect, white,
and urn shaped, and bloom in drooping, curved racemes up to 20 cm
(7.9 in) long in summer. Fruit is a woody capsule, 3–8 mm (0.1–0.3
in) long, and clustered on curved or drooping stalks; fruit matures in

Clockwise from upper left:

Sourwood leaves.

Sourwood fruit.

Sourwood flowers.

Clockwise from upper left:

Sourwood form and habitat.

Sourwood bark.

Sourwood bark with orange in the grooves.

early fall but persists over the winter. Bark is gray-black and deeply grooved with orange-red within the grooves; on large trees the bark is blockier. The trunk is often crooked. The growth form is up to 24 m (80 ft) in height and 60 cm (23.6 in) in diameter.

HABITAT Slopes, ridges, and a variety of well-drained sites.

NOTES Sourwood can be found growing with upland hickories, pines, and oaks. The wood is light yellow-brown to reddish-brown, heavy, and hard, and is used for pulpwood and specialty items such as tool handles. Sourwood is an attractive landscape tree because of the flowers, fall color, and stress tolerance. It is a browse for white-tailed deer, and the flowers are an important source of nectar for honey bees (sourwood honey). The sour tasting leaves are used in teas.

Oxydendrum means "sour tree"; *arboreum* means "treelike."

Sparkleberry

Vaccinium arboreum Marshall

COMMON NAMES sparkleberry, farkleberry, tree huckleberry

QUICK GUIDE Leaves alternate, simple, small, glossy, margin entire or with small teeth; fruit a berry, blue-black, juicy; bark red-brown and shreddy.

DESCRIPTION Leaves are alternate, simple, tardily deciduous, oval to elliptical, 2–7 cm (0.8–2.8 in) long, dark green, shiny, and nearly sessile; apex is sometimes mucronate; margin is entire or with small glandular teeth; autumn color is red. Twigs are slender and red-brown; leaf scar is shield shaped with one bundle scar. A true terminal bud is lacking; lateral bud is round, small (about 1 mm [0.1 in] long), and embedded. Flowers are perfect, white, and bell shaped, and bloom in attractive racemes in the spring after the leaves. Fruit is a berry, round, 6 mm (0.2 in) wide, red-blue to blue-black, and juicy, and matures in the summer or early fall. Bark is red-brown to gray and peels to reveal red inner bark. The growth form is a shrub or small twisted tree up to 10 m (30 ft) tall.

HABITAT A variety of sites and well-drained soils, including sand dunes, dry slopes, forest edges, and stream margins.

Clockwise from left:

Sparkleberry leaves.

Sparkleberry flowers.

Sparkleberry immature fruit.

Sparkleberry bark.

NOTES Sparkleberry wood is close-grained and hard, and is used for woodenware, smoking pipes, and walking canes. The fruit is eaten by a variety of birds and mammals, but they prefer other *Vaccinium* species because it tends to be gritty. *Vaccinium* is Latin for "blueberry"; *arboreum* means "treelike."

Spurge Family (*Euphorbiaceae*)

Chinese Tallowtree

Triadica sebifera (L.) Small

COMMON NAMES Chinese tallowtree, popcorn tree

QUICK GUIDE Leaves alternate, simple, rhombic,
apex long and tapered, petiole with glands at the
blade; flowers in drooping bright yellow clusters in the spring;
fruit a popcornlike capsule in the fall; bark gray-brown with
orange-red in the grooves.

DESCRIPTION Leaves are alternate, simple, deciduous, rhombic or deltoid, and 5–8 cm (2.0–3.1 in) long; apex tapers to a very long point;
margin is entire; petiole has a pair of glands at the blade; autumn
color is yellow, orange, or red. Twigs are green-brown and glaucous,
with lenticels; leaf scar is shield shaped with three bundle scars; sap is
milky and poisonous. Terminal bud is triangular and naked. Flowers
are imperfect and in bright yellow, drooping spikes up to 20 cm (7.9
in) long in the spring after the leaves; the spike is mostly staminate
flowers with a few pistillate flowers at the base. Fruit is a three-lobed

Clockwise from left:

Chinese tallowtree
leaves and flowers.

Chinese tallowtree
leaves and flowers.

Chinese tallowtree
fruit.

Chinese tallowtree
bark.

capsule that contains three white, waxy seeds about 1 cm (0.4 in) long. The seeds resemble popcorn kernels and the fruit matures in the fall. Bark is gray-brown and grooved, with interlacing ridges and orange-red between the ridges. The growth form is up to 15 m (50 ft) in height.

HABITAT A variety of sites.

NOTES Chinese tallowtree can form pure stands or thickets and is an aggressive colonizer of open sites. It is a native of eastern China and listed as legally noxious and considered a harmful invasive species in many states in the Southeast. It is a direct competitor of many native plants. In China, the wax from the seed coat was used in making candles.

Triadica means three parted, referring to the fruit; *sebifera* means "tallow," also referring to the fruit. Some texts use *Sapium sebiferum* (L.) Small.

Bean or Pea Family (*Fabaceae*)

Mimosa

Albizia julibrissin Durazz.

COMMON NAMES mimosa, silk tree

QUICK GUIDE Leaves alternate, bipinnately compound, leaflets small and lopsided, midrib close to and parallel to one edge; flowers resemble pink pom-poms, fragrant, blooming throughout the summer; fruit a legume; bark gray and smooth.

DESCRIPTION Leaves are alternate, bipinnately compound, deciduous, and up to 45 cm (17.7 in) long. Leaflets are about 1 cm (0.4 in) long, and appear lopsided; midrib is close to and parallel to one edge; margin is entire; autumn color is yellow. Twigs are stout, mottled gray-brown to black, and glabrous, with lenticels; leaf scar is raised and three-lobed, with three bundle scars. A true terminal bud is lacking; lateral bud is minute and round; scales are overlapping, brown, and glabrous. Flowers are perfect and very fragrant, and have silky, bright pink filaments clustered in erect pom-poms. The flowers bloom

Clockwise from left:

Mimosa leaf.

Mimosa flowers.

Mimosa leaves, flowers, and fruit.

Mimosa bark.

throughout the summer. Fruit is a brown legume up to 20 cm (7.9 in) long and matures in the fall. Bark is pale gray and smooth. The growth form is usually only up to 9 m (30 ft) in height with a flat top and low, open crown.

HABITAT Open sites such as in fields and along fencerows, roads, and wood edges.

NOTES Mimosa is originally from China and was commonly planted in the South as an ornamental. This species has now become naturalized throughout the state and is considered an invasive plant.

Albizia is for an Italian naturalist, Filippo Albizzi, who brought this tree to Italy; *julibrissin* comes from the Persian name meaning floss silk and refers to the silky flowers.

Eastern Redbud

Cercis canadensis L.

COMMON NAMES eastern redbud, Judas-tree, redbud

QUICK GUIDE Leaves alternate, simple, heart shaped, palmately veined, glabrous, margin entire, petiole swollen at both ends; flowers in pink-purple clusters along branches in early spring; fruit podlike; bark red-brown to black, shallowly ridged or scaly.

DESCRIPTION Leaves are alternate, simple, deciduous, kidney or heart shaped, and palmately veined; blade is 8–12 cm (3.1–4.7 in) long; apex is abruptly acute; base is cordate; margin is entire; petiole is up to 9 cm (3.5 in) long and swollen at both ends; underside is glabrous; autumn color is yellow. Twigs zigzag and are slender, red-brown to black, and glabrous, with lenticels; leaf scar is heart shaped and raised with three bundle scars. A true terminal bud is lacking; lateral bud is brown to black, about 2 mm (0.1 in) long, triangular, and appressed; flower bud is plump and sits above the leaf bud. Flowers are perfect, about 1 cm (0.4 in) long, and rose pink to purplish, and cluster along the branch before the leaves. Fruit is a flat legume, up to 10 cm (3.9 in) long, and red-brown to black and matures in late summer. Bark is red-brown to black and smooth on small trees and shallowly ridged or scaly on large trees. The growth form is an understory tree up to 12 m (40 ft) in height with a low, open crown.

Clockwise from upper left:

Eastern redbud leaf.

Eastern redbud fruit.

Eastern redbud flowers.

Eastern redbud
bark.

HABITAT In the understory on rich moist soils especially near streams but also found on drier upland sites.

NOTES Eastern redbud is a shade-tolerant tree found growing in the forest understory with a wide variety of species. The wood is of little commercial value. The seed is eaten by birds (including northern bobwhite), white-tailed deer, and small mammals. The flowers are visited by bees. Eastern redbud is a very popular ornamental because of the early spring flowers that bloom even on small trees and its tolerance of a range of conditions. Cultivars have more deeply colored flowers or white flowers and purple foliage.

Cercis is the classic Greek name for the Judas-tree of southern Europe and Asia; *canadensis* refers to the New World.

Honeylocust

Gleditsia triacanthos L.

COMMON NAMES honeylocust, thorny-locust, sweet-locust

QUICK GUIDE Leaves alternate, pinnately or bipinnately compound, with up to 30 leaflets; twigs with branched thorns; fruit podlike, long, often twisted; bark with clumps of thorns, gray-brown, smooth or with large scaly plates.

DESCRIPTION Leaves are alternate, pinnately or bipinnately compound, deciduous, and up to 20 cm (7.9 in) long, and have 14 to 30 leaflets. Leaflets are elliptical, about 2 cm (0.8 in) long, dark green, shiny, and sessile; margin is entire or lightly toothed; underside is pubescent; leaflets subopposite to alternate; autumn color is yellow. Twigs zigzag and are stout, green to red-brown, and glabrous, with lenticels and branched thorns; leaf scar is heart shaped with three bundle scars. A true terminal bud is lacking; lateral bud is almost completely embedded and sunken. Flowers are perfect and imperfect with five yellow-green petals and greatly exserted stamens, and they bloom in dangling racemes after the leaves. Fruit is a flat legume, up to 45 cm (17.7 in) long, red-brown, and often twisted or curved; pulp around the kidney-shaped, flattened seeds is very sweet when the pod is yellow-brown. Bark is gray-brown and smooth, often with dense clusters of thorns on small trees; on large trees the bark has large, loose plates. The growth form is up to 24 m (80 ft) in height and 1 m (3 ft) in diameter.

HABITAT Naturalized throughout the state and found on a variety of sites but commonly near streams and in bottomlands.

From left to right:

Honeylocust leaves.

Honeylocust flowers.

Honeylocust fruit.

From left to right:

Bark of a small honeylocust with thorns.

Bark of large honeylocust with scaly plates.

NOTES Honeylocust is found growing with a variety of species. The wood is red-brown to yellow, very heavy, very hard, and durable, and is used for fence posts, furniture, and specialty items. The wood was used for wheel hubs. The fruit is eaten by birds (including northern bobwhite), small mammals, and cattle. Entire pods are relished by white-tailed deer. The flowers are popular with bees. Thornless cultivars with enhanced disease and pollution resistance are commonly planted in cities and towns.

Gleditsia is in honor of the botanist Johann Gleditsch; *triacanthos* refers to the three-branched thorns. Honeylocust is one word because it is not a true locust.

SIMILAR SPECIES Waterlocust (*Gleditisia aquatica* Marshall) is similar to honeylocust but is found on wetter sites (in swamps, hammocks, and floodplains) and is distinguished by a kidney-shaped flat legume up to 5 cm (2 in) long that contains one to three seeds and lacks pulp around the seeds.

Black Locust

Robinia pseudoacacia L.

COMMON NAMES black locust, yellow locust, yellow ash

QUICK GUIDE Leaves alternate, pinnately compound, with 7 to 19 leaflets, margin entire; twigs with a stout pair of spines at each node; flowers white, fragrant, in pendent clusters; fruit podlike; bark light brown to gray with thick interlacing ridges.

DESCRIPTION Leaves are alternate, pinnately compound, deciduous, and 20–36 cm (7.9–14.2 in) long, and have 7 to 19 leaflets. Leaflets are oblong or oval and 2–5 cm (0.8–2.0 in) long; apex is rounded to emarginate; base is obtuse to round; margin is entire; autumn color is yellow. Twigs zigzag and are moderately stout, angular, and red-brown, with a pair of stout spines at the node; leaf scar is triangular to three-lobed with three bundle scars. A true terminal bud is lacking; lateral bud is minute, naked, embedded, and hidden under the leaf scar. Flowers are perfect, white, and fragrant, and bloom in pendent racemes in the spring after the leaves. Fruit is a legume, up to 10 cm (3.9 in) long, flat, and brown, and matures in the fall. Bark is light brown to gray and furrowed with thick, interlacing corky or scaly

From left to right:

Black locust leaf.

Black locust flowers and fruit.

From left to right:

Black locust bark.

Obovate leaflets of
a yellowwood leaf
for comparison.

ridges. The growth form is up to 21 m (70 ft) in height and 61 cm (24 in) in diameter and can form thickets.

HABITAT Grows best on moist limestone soils but is found on a variety of sites ranging from moist slopes and stream banks to dry rocky soils and old fields, often coming in after a disturbance. Can sprout from the roots and stump.

NOTES Forest associates of black locust include numerous upland oaks, pines, and hickories. The very durable wood is yellow-green to red-brown, very hard, and very heavy, and was once used for wooden nails and in shipbuilding. Modern uses include pulpwood, boxes, woodenware, fuel, fence posts, and railroad ties. Black locust is used in waste site reclamation because of its ability to fix nitrogen. The fruit is eaten by northern bobwhite, mourning dove, foxes, opossum, and white-tailed deer. The bark and seeds are reported to be poisonous to livestock and humans. Black locust flowers are a favorite of honey bees.

Robinia is for the French herbalist Jean Robin; *pseudoacacia* refers to the leaves and spines, which are similar to the genus *Acacia* ("akakia" means Egyptian thorn tree).

SIMILAR SPECIES **Yellowwood** (*Cladrastis kentukea* [Dum.Cours.] Rudd) is in the same family, has similar flowers, and is found on rich moist soils in a few counties in Alabama. It is distinguished by alternate, pinnately compound leaves with 7 to 11 obovate leaflets; twigs with no spines and a leaf scar nearly encircling the gold fuzzy bud; a constricted legume up to 10 cm (3.9 in) long; and very smooth bark even on large trees.

Beech Family (*Fagaceae*)

American Chestnut

Castanea dentata (Marshall) Borkh.

COMMON NAMES American chestnut

QUICK GUIDE Leaves alternate, simple, elliptical, margin with coarse and curved bristle-tipped teeth; twigs and buds glabrous; fruit two to three triangular nuts enclosed in a large spiny bur.

DESCRIPTION Leaves are alternate, simple, deciduous, oblong to elliptical or lanceolate, and 13–20 cm (5.1–7.9 in) long; apex is acuminate; base is cuneate to rounded; margin has coarse and curved bristle-tipped teeth; petiole is glabrous; underside is mostly glabrous; autumn color is yellow. Twigs are slender, red-brown to brown-gray, and glabrous; leaf scar is crescent shaped to oval with numerous bundle scars. A true terminal bud is lacking; lateral bud is ovoid, about 6 mm (0.2 in) long, and chestnut brown to orange-brown, with two to three glabrous, overlapping scales. Flowers are imperfect and appear after the leaves; staminate catkins are long and stiff; pistillate flowers are at the base of staminate catkins or from leaf axils. Fruit is a nut, nearly round, up to 3 cm (1.2 in) wide, flattened, and shiny brown. Two to three nuts are enclosed in a spiny bur that matures in the fall. Bark is red-brown to gray-brown or black and smooth or ridged on small trees; on large trees the bark is gray-brown to dark brown with thick, broad, interlacing ridges. The growth form was once a large tree up to 30 m (100 ft)

From left to right:

American chestnut leaves.

American chestnut fruit.

From left to right:

Bark of an American chestnut sapling.

Chinese chestnut leaf with smaller margin teeth for comparison.

in height, but because of chestnut blight this species currently exists only as sprouts or small trees up to about 9 m (30 ft) in height.

HABITAT Rich soils and upland sites, now uncommon due to chestnut blight.

NOTES Before the chestnut blight, majestic (more than 3 m [10 ft] in diameter) American chestnut trees once dominated the hardwood forests of the Appalachian Mountains. The blight survives on oaks and infects American chestnut trees when the bark becomes furrowed. Usually only sprouts can be found, but trees with fruit have been reported. Hybrids with Chinese chestnut (see below) that are resistant to the blight are being tested in order to save the species. This tree was commercially important before the blight, and the hard, durable red-brown wood was used for a range of products such as construction lumber, flooring, paneling, furniture, caskets, and trim work. The nuts were an important food source for wild turkey, foxes, white-tailed deer, other small and large mammals, songbirds, and people. The leaves were used in medicinal teas and salves.

Castanea is Latin for "chestnut tree"; *dentata* means "tooth," referring to the leaves.

SIMILAR SPECIES Chinese chestnut (*Castanea mollissima* Blume) has been introduced in the state and is identified by leathery leaves with smaller teeth on the margin, white tomentose hairs on the underside, buds with white hairs, and a larger bur (up to 7 cm [2.8 in] wide) with two to three nuts enclosed. Chinese chestnut is planted for shade and its edible nuts and is often found at old home sites but the male flowers are foul smelling and the burs are very spiny.

Allegheny Chinkapin

Castanea pumila (L.) Mill.

COMMON NAMES Allegheny chinkapin, chinquapin

QUICK GUIDE Leaves alternate, simple, elliptical, margin with coarse bristle-tipped teeth, underside pubescent to tomentose; twigs tomentose; fruit a round nut enclosed in a spiny bur; bark red-brown to gray, becomes fissured.

DESCRIPTION Leaves are alternate, simple, deciduous, oblong to elliptical or oblanceolate, and 8–15 cm (3.1–5.9 in) long; apex is rounded to acute; base is cuneate to rounded; margin has coarse, bristle-tipped teeth; underside and petiole are pubescent to tomentose; autumn color is yellow-brown. Twigs are slender, red-brown to gray, and pubescent to tomentose; leaf scar is crescent shaped to oval with numerous bundle scars. The terminal bud is ovoid to round, and about 4 mm (0.2 in) long, with two to three red-brown, hairy scales. Flowers are imperfect and appear after the leaves; staminate catkins are long and stiff; pistillate flowers are on the base of staminate catkins or from leaf axils. Fruit is a nut, ovoid, up to 2 cm (0.8 in) wide, and shiny brown, and one nut is enclosed in a spiny bur. The fruit matures in the fall. Bark is red-brown to gray and smooth on small trees; large trees have shallowly fissured bark. The growth form is a shrub or tree up to 6 m (20 ft) in height.

Clockwise from left:

Allegheny chinkapin leaves.

Allegheny chinkapin leaves and developing fruit; courtesy of Nancy Loewenstein.

Allegheny chinkapin fruit; courtesy of Alan Cressler.

Allegheny chinkapin bark; courtesy of Nancy Loewenstein.

HABITAT Sandy ridges and upland forests.

NOTES Forest associates of Allegheny chinkapin include pignut hickory, mockernut hickory, eastern redcedar, shortleaf pine, scarlet oak, southern red oak, chestnut oak, black oak, and winged elm. This species is susceptible to chestnut blight. The wood was used for fences and railroad ties. The nuts are eaten by wild turkey, foxes, white-tailed deer, other small and medium-size mammals, songbirds, and people. *Castanea* is Latin for "chestnut tree"; *pumila* means dwarf, relative to American chestnut.

American Beech

Fagus grandifolia Ehrh.

COMMON NAMES American beech, beechnut tree

QUICK GUIDE Leaves alternate, simple, elliptical, each lateral vein ending at a margin tooth; buds large, long-pointed, cigarlike; fruit a triangular nut, with usually two enclosed in a weakly spiny bur; bark smooth and blue-gray.

DESCRIPTION Leaves are alternate, simple, deciduous, elliptical, and 8–13 cm (3.1–5.1 in) long, at first very thin and pubescent but become more leathery, shiny, and glabrous throughout the summer; lateral veins are parallel and each vein ends at a margin tooth; apex is acuminate to rounded; base is obtuse; margin is sharply serrate; autumn color is orange-yellow, and dead leaves persist throughout the winter. Twigs zigzag and are slender, yellow-brown to red-brown, and glabrous, with lenticels; leaf scar is crescent shaped to half-round with three or more bundle scars. The terminal bud is cigarlike, up to 2.5 cm (1 in) long, and long-pointed; scales overlap and are shiny and yellow-brown to red-brown. Flowers are imperfect and bloom in the spring with the leaves; staminate flowers grow in heads and are long-stalked and drooping; pistillate spikes are much smaller and grow in the leaf axils at the shoot tip. Fruit is a nut, about 1.5 cm (0.6 in) long, triangular, and sweet and matures in the fall. Usually two nuts are enclosed in a woody and weakly spiny bur. Bark is blue-gray, thin, and

From left to right:

American beech leaves and fruit.

American beech flowers.

American beech fruit.

From left to right:

American beech bark.

American beech tree with winter leaves; photo by Lisa J. Samuelson.

smooth. The growth form is up to 30 m (100 ft) in height and 1.3 m (4 ft) in diameter with a low, wide crown.

HABITAT Fertile, moist, loamy soils of bottomlands, coves, stream edges, and north-facing slopes.

NOTES American beech is a shade-tolerant, long-lived tree found growing with red maple, sugar maple, black cherry, tulip-poplar, southern magnolia, sweetbay magnolia, eastern white pine, many mesic site oak species, basswood, and American elm. The wood is yellow-brown, heavy, and hard with tyloses in the heartwood and distinct growth rings. It is used for furniture (chairs in particular), veneer, flooring, tool handles, boxes, and turnery. The nuts are relished by wild turkey, foxes, wood duck, numerous songbirds, black bear, squirrels, and various other small mammals. This tree would be even more valuable to wildlife, but it has frequent seed crop failures.

Fagus is Latin for "beech tree," which comes from a Greek word for "to eat"; *grandifolia* means "large leaved."

White Oak

Quercus alba L.

COMMON NAMES white oak, stave oak

QUICK GUIDE Leaves alternate, simple, with seven to nine rounded and bristleless lobes; acorn cap knobby; bark gray-white, loosely plated or scaly.

DESCRIPTION Leaves are alternate, simple, deciduous, and 10–18 cm (3.9–7.1 in) long, with seven to nine rounded and bristleless lobes; apex is rounded; sinuses extend halfway or almost to the midrib; base is cuneate; underside is pale and glabrous or glaucous; autumn color is dull red or orange-yellow. Twigs are moderately stout, red-brown, and glabrous when young; leaf scar is crescent shaped to oval with numerous bundle scars. Terminal buds are about 4 mm (0.2 in) long and ovoid to round; scales are overlapping, red-brown, and mostly glabrous. Flowers are imperfect and appear in the spring with the leaves; staminate flowers are in drooping catkins; pistillate spikes are in leaf axils. Fruit is an acorn, 2–3 cm (0.8–1.2 in) long; nut is shiny and yellow to light brown; cap is bowl shaped with knobby, glabrous scales covering one-fourth to one-third of the nut; fruit matures in one season. Bark is gray-white to gray-brown and shallowly grooved, with small rectangular scales or large loose plates; bark becomes deeply grooved at the base of large trees. The growth form is commonly 30 m (100 ft) in height and 1.3 m (4 ft) in diameter.

HABITAT A variety of sites, including ridges, coves, sandy plains, dry slopes, and bottomlands.

From left to right:

White oak leaf.

White oak flowers.

White oak fruit, note the knobby cap of the acorn.

From left to right:

White oak bark with loose plates.

White oak grooved bark.

NOTES Forest associates of white oak are numerous and site dependent. "White oak lumber" is yellow-brown, heavy, and hard, with tyloses, and is an important commercial wood used for flooring, furniture, trim work, and staves for barrels. The bark was used in tanneries. White oak is a nice landscape tree for large spaces because of its bark, foliage, and form. Acorns of all oaks rank at the top of wildlife food plants because of their wide distribution, availability, and palatability, and are considered the staff of life for many wildlife species. Acorns of the white oak group are generally preferred over the red oak group because of lower tannic acid levels. Smaller birds and mammals prefer the species producing the smaller acorns, whereas larger birds and mammals will eat all sizes, including the largest. A partial list of acorn eaters include mallard, black, pintail, and wood ducks; northern bobwhite; wild turkey; flying, fox, and gray squirrels; all species of deer; black bear; peccary; raccoon; numerous songbirds; and many small mammals.

Quercus is Latin for "oak tree"; *alba* means "white," perhaps referring to the bark or underside of leaves.

Bluff Oak

Quercus austrina Small

COMMON NAMES bluff oak

QUICK GUIDE Leaves alternate, simple, with three to seven rounded and bristleless lobes, lobes and sinuses variable in shape; acorn cap with thin scales; bark gray-white and scaly.

DESCRIPTION Leaves are alternate, simple, deciduous, and 5–15 cm (2.0–5.9 in) long, with three to seven bristleless lobes but some leaves may be unlobed; lobes and sinuses are variable in shape; apex is rounded to acute; base is cuneate; underside is glabrous; autumn color is dull red or orange-yellow. Twigs are red-brown and pubescent or glabrous; leaf scar is crescent shaped to oval with numerous bundle scars. Terminal buds are ovoid and about 4 mm (0.2 in) long; scales are overlapping, red-brown, and pubescent. Flowers are imperfect and emerge with the leaves in the spring; staminate flowers are in drooping catkins; pistillate spikes are in leaf axils. Fruit is an acorn about 2 cm (0.8 in) long; cap has thin, pubescent scales covering one-fourth to one-third of the nut; fruit matures in one season. Bark is gray-white to gray-brown and loosely ridged and becomes scaly on large trees. The growth form is up to 23 m (75 ft) in height.

Clockwise from left:

Bluff oak leaves with little lobing.

Bluff oak fruit.

Bluff oak leaves with lobes.

Bluff oak bark.

HABITAT An occasional tree found on rich soils of stream edges and river bluffs and on shell or calcareous sediments.

NOTES Forest associates of bluff oak include devilwood, redbay, cherry laurel, laurel oak, mountain laurel, sparkleberry, and hophornbeam. Bluff oak is used as white oak lumber. Acorns of all oaks rank at the top of wildlife food plants because of their wide distribution, availability, and palatability, and are considered the staff of life for many wildlife species. Acorns of the white oak group are generally preferred over the red oak group because of lower tannic acid levels. Smaller birds and mammals prefer the species producing the smaller acorns, whereas larger birds and mammals will eat all sizes, including the largest. A partial list of acorn eaters include mallard, black, pintail, and wood ducks; northern bobwhite; wild turkey; flying, fox, and gray squirrels; all species of deer; black bear; peccary; raccoon; numerous songbirds; and many small mammals.

Quercus is Latin for "oak tree"; *austrina* means "south."

Scarlet Oak

Quercus coccinea Muenchh.

COMMON NAMES scarlet oak, red oak, Spanish oak

QUICK GUIDE Leaves alternate, simple, with five to nine bristle-tipped lobes, sinuses deep; buds with pubescence on the upper half; acorn often with concentric grooves on the nut apex; bark gray with flat ridges or dark and rough with white streaks.

DESCRIPTION Leaves are alternate, simple, deciduous, and 10–20 cm (3.9–7.9 in) long, with five to nine bristle-tipped lobes; sinuses extend more than halfway to the midrib and are deepest in sun leaves; base is truncate to acute; vein axils have tufts of hair; autumn color is scarlet. Twigs are red-brown to gray-brown and glabrous; leaf scar is crescent shaped to oval with numerous bundle scars. Terminal buds are about 6 mm (0.2 in) long, acute, and slightly angled; scales are overlapping and brown, with white or light brown pubescence on the upper half of the bud. Flowers are imperfect and emerge in the spring with the leaves; staminate flowers are in drooping catkins; pistillate spikes are in leaf axils. Fruit is an acorn, 1.3–2.5 cm (0.5–1.0 in) long; nut is pubescent and may have concentric thin grooves around the apex; cap is bowl-like with shiny appressed scales covering one-third to one-half of the nut; fruit matures in two seasons. Bark is gray with flat ridges and can be gray-black and rough at the base and have white streaks on the middle and upper trunk; inner bark is orange-brown. Branches are often nearly horizontal, and the trunk may show butt swell. Dead

From left to right:

Scarlet oak shade leaf.

Scarlet oak fruit, note the grooves at the nut apex.

From left to right:

Scarlet oak bark with white streaks.

Scarlet oak bark with flat ridges.

Leaf (above) and striped acorn (below) of pin oak for comparison.

branches may persist in the lower canopy. The growth form is up to 24 m (80 ft) in height and 1 m (3 ft) in diameter.

HABITAT A variety of sites but common on dry, lightly sandy, or rocky soils.

NOTES Scarlet oak is found in mixed stands with blackgum, sourwood, shortleaf pine, loblolly pine, Virginia pine, and numerous other upland oaks. The wood is used as red oak lumber and for trim, flooring, and furniture. Scarlet oak is planted as an ornamental for its fall color and site adaptability. Acorns of all oaks rank at the top of wildlife food plants because of their wide distribution, availability, and palatability, and are considered the staff of life for many wildlife species. Acorns of the white oak group are generally preferred over the red oak group because of lower tannic acid levels. Smaller birds and mammals prefer the species producing the smaller acorns, whereas larger birds and mammals will eat all sizes, including the largest. A partial list of acorn eaters include mallard, black, pintail, and wood ducks; northern bobwhite; wild turkey; flying, fox, and gray squirrels; all species of deer; black bear; peccary; raccoon; numerous songbirds; and many small mammals.

Quercus is Latin for "oak tree"; *coccinea* means "scarlet," referring to the autumn leaves.

SIMILAR SPECIES Pin oak (*Quercus palustris* Muenchh.) is rare in Alabama but has been reported in mesic woods in northern Alabama and is a common ornamental tree throughout the state. It is distinguished by leaves with five to nine bristle-tipped lobes and deep U- or C-shaped sinuses, smooth or lightly ridged bark, a maroon-striped nut, and branches that often form 90-degree angles to the stem.

Durand Oak

Quercus durandii Buckley

COMMON NAMES Durand oak, Durand white oak

QUICK GUIDE Leaves alternate, simple, spatulate, unlobed or shallowly lobed near the apex, margin wavy and lacking bristle tips, underside densely hairy; acorn cap covering the base of the nut; bark gray-brown and scaly.

DESCRIPTION Leaves are alternate, simple, tardily deciduous, obovate to spatulate, 12–18 cm (4.7–7.1 in) long, dark green, and shiny; apex is rounded and lacks a bristle tip; base is cuneate; margin is sinuate and unlobed or shallowly lobed at the apex; midrib and petiole are yellow; underside has dense stellate pubescence. Twigs are gray-brown and glabrous; leaf scar is crescent shaped to oval with numerous bundle scars. Terminal buds are about 5 mm (0.2 in) long and ovoid; scales are overlapping, red-brown, and glabrous or lightly pubescent. Flowers are imperfect and appear in the spring with the leaves; staminate flowers are in drooping catkins; pistillate spikes are in leaf axils. Fruit is an acorn, 1.3–2.0 cm (0.5–0.8 in) long; nut is brown and flat topped; cap is saucer shaped with pubescent scales covering only the base of the nut; fruit matures in one season. Bark is gray-brown and scaly. The growth form is up to 27 m (90 ft) in height.

HABITAT An occasional tree found on limestone soils and on the rich well-drained soils of bottomlands and slopes.

Durand oak leaves.

Durand oak bark.

NOTES Durand oak is uncommon and found with hackberry, nutmeg hickory, bitternut hickory, American elm, white ash, willow oak, Shumard oak, and live oak. It is used as white oak lumber. Acorns of all oaks rank at the top of wildlife food plants because of their wide distribution, availability, and palatability, and are considered the staff of life for many wildlife species. Acorns of the white oak group are generally preferred over the red oak group because of lower tannic acid levels. Smaller birds and mammals prefer the species producing the smaller acorns, whereas larger birds and mammals will eat all sizes, including the largest. A partial list of acorn eaters include mallard, black, pintail, and wood ducks; northern bobwhite; wild turkey; flying, fox, and gray squirrels; all species of deer; black bear; peccary; raccoon; numerous songbirds; and many small mammals.

Quercus is Latin for "oak tree"; *durandii* is for the botanist Elias Durand. Also referred to as *Quercus sinuata* Walter *var. sinuata.*

Southern Red Oak

Quercus falcata Michx.

COMMON NAMES southern red oak, Spanish oak

QUICK GUIDE Leaves alternate, simple, drooping, with three to seven bristle-tipped lobes, terminal lobe elongated, base bell shaped, underside densely hairy; acorn cap covering up to half of the pubescent nut; bark dark and rough.

DESCRIPTION Leaves are alternate, simple, deciduous, 10–23 cm (3.9–9.1 in) long, and drooping with three to seven bristle-tipped lobes; terminal lobe is elongated and sickle shaped; sinuses are long and irregular; base is bell shaped to obtuse and more likely to be obtuse on juvenile trees; underside has dense yellow or rusty pubescence or tomentum; autumn color is yellow-brown. Three-lobed leaves are common on seedlings. Twigs are gray-brown and pubescent; leaf scar is crescent shaped to oval with numerous bundle scars. Terminal buds are about 6 mm (0.2 in) long, ovoid to acute, and slightly angled; scales are overlapping, red-brown, and pubescent. Flowers are imperfect and appear in the spring with the leaves; staminate flowers are in drooping catkins; pistillate spikes are in leaf axils. Fruit is an acorn, about 1.3 cm (0.5 in) long; nut is pubescent and sometimes striped; cap is bowl-like with appressed pubescent scales covering one-third to one-half of the nut; fruit matures in two seasons. Bark is dark gray

From left to right:

Southern red oak leaves.

Southern red oak fruit.

to black and rough on small trees and deeply grooved and ridged on large trees; inner bark is pale yellow. The growth form is up to 25 m (80 ft) in height and 1 m (3 ft) in diameter.

HABITAT Dry, upland forests.

NOTES Southern red oak is found growing with blackgum, sourwood, shortleaf pine, loblolly pine, Virginia pine, longleaf pine, pignut and mockernut hickory, and numerous other upland oaks. It is used as red oak lumber for construction and furniture. Acorns of all oaks rank at the top of wildlife food plants because of their wide distribution, availability, and palatability, and are considered the staff of life for many wildlife species. Acorns of the white oak group are generally preferred over the red oak group because of lower tannic acid levels. Smaller birds and mammals prefer the species producing the smaller acorns, whereas larger birds and mammals will eat all sizes, including the largest. A partial list of acorn eaters include mallard, black, pintail, and wood ducks; northern bobwhite; wild turkey; flying, fox, and gray squirrels; all species of deer; black bear; peccary; raccoon; numerous songbirds; and many small mammals.

Quercus is Latin for oak "tree"; *falcata* means "scythe or sickle shaped," referring to the leaves.

Laurel Oak

Quercus hemisphaerica Bartram ex Willd.

COMMON NAMES laurel oak, sand laurel oak, Darlington oak

QUICK GUIDE Leaves alternate, simple, elliptical or oblanceolate, unlobed, margin entire or occasionally toothed, apex bristle tipped, vein axils mostly glabrous; acorn cap covering the base of the small nut; bark gray, smooth or shallowly ridged.

DESCRIPTION Leaves are alternate, simple, evergreen or tardily deciduous, elliptical to oblanceolate or sometimes obovate, 5–12 cm (2.0–4.7 in) long, leathery, shiny, and nearly sessile; apex is acute with a bristle tip; base is acute to cuneate; margin is usually entire but sometimes has teeth or small lobes, especially on young trees; midrib is yellow and mostly glabrous. Twigs are slender, gray-brown, and glabrous; leaf scar is crescent shaped to oval, with numerous bundle scars. Terminal buds are about 3 mm (0.1 in) long and ovoid; scales are overlapping and red-brown with pubescence on the margins. Flowers are imperfect and appear in the spring with the leaves; staminate flowers are in drooping catkins; pistillate spikes are in leaf axils. Fruit is an acorn, about 1.3 cm (0.5 in) long; nut is brown, flat topped, and pubescent; cap is flat with appressed pubescent scales covering one-fourth or less of the nut; fruit matures in two seasons. Bark is brown-gray to gray-black and smooth or shallowly ridged on young trees; very large trees have scaly ridges. The growth form is up to 24 m (80 ft) in height.

HABITAT Well-drained, sandy soils near streams, rivers, and swamp edges, and in dry uplands.

From left to right:
Laurel oak leaves.
Laurel oak fruit.

Clockwise from left:

Bark of a young laurel oak.

Bark of a large laurel oak.

Laurel oak seedling with lobed leaves.

Shingle oak leaves, which are longer with a pubescent to tomentose underside for comparison.

NOTES Laurel oak is found with a variety of species depending on the site. The wood is used as red oak lumber and for pulpwood and fuel. Laurel oak is planted as an ornamental tree because of its persistent, shiny leaves and is often crossed with willow oak. Acorns of all oaks rank at the top of wildlife food plants because of their wide distribution, availability, and palatability, and are considered the staff of life for many wildlife species. Acorns of the white oak group are generally preferred over the red oak group because of lower tannic acid levels. Smaller birds and mammals prefer the species producing the smaller acorns, whereas larger birds and mammals will eat all sizes, including the largest. A partial list of acorn eaters include mallard, black, pintail, and wood ducks; northern bobwhite; wild turkey; flying, fox, and gray squirrels; all species of deer; black bear; peccary; raccoon; numerous songbirds; and many small mammals.

Quercus is Latin for "oak tree"; *hemisphaerica* refers to the shape of the nut. Confusion exists in naming this species. Some texts refer to this species as swamp laurel oak (*Quercus laurifolia*).

SIMILAR SPECIES Shingle oak (*Quercus imbricaria* Michx.) is similar to laurel oak but has a limited distribution in Alabama. It is identified by elliptical to oblong or oblanceolate leaves, 8–18 cm (3.1–7.1 in) long, with an underside that is pubescent to tomentose. The acorn is about 2 cm (0.8 in) long, and the cap has pubescent appressed scales and covers one-third to one-half of the pubescent nut.

Bluejack Oak

Quercus incana Bartram

COMMON NAMES bluejack oak, sand jack oak, scrub oak

QUICK GUIDE Leaves alternate, simple, oblong or oblanceolate, margin entire, apex bristle tipped, underside with blue-gray pubescence; acorn cap covering up to one-third of the small nut; bark dark, rough, and blocky.

DESCRIPTION Leaves are alternate, simple, deciduous, oblong to elliptical or oblanceolate to lanceolate, 5–12 cm (2.0–4.7 in) long, leathery, and nearly sessile; apex is obtuse with a bristle tip; base is acute; margin is entire or has an occasional tooth or small lobe, mostly on seedlings; upperside is shiny and blue-green; underside has prominent blue-gray or white pubescence; autumn color is yellow-brown. Twigs are slender and gray-brown, with blue-gray pubescence; leaf scar is crescent shaped to oval with numerous bundle scars. Terminal buds are about 5 mm (0.2 in) long and acute; scales are overlapping, red-brown, and lightly pubescent. Flowers are imperfect and appear in the spring with the leaves; staminate flowers are in drooping catkins; pistillate spikes are in leaf axils. Fruit is an acorn, about 1 cm (0.4 in) long; nut is pubescent; cap is bowl shaped with pubescent, appressed scales covering one-fourth to one-third of the nut; fruit matures in two seasons. Bark is gray-black, rough, and blocky. The growth form is a small tree up to 9 m (30 ft) tall.

From left to right:

Bluejack oak leaves.

Bluejack oak fruit.

From left to right:

Bark of a large bluejack oak tree; photo by Lisa J. Samuelson.

Revolute margins of myrtle oak leaves for comparison.

HABITAT Well-drained, dry, and sandy soils.

NOTES Bluejack oak is found growing with sand hickory, longleaf pine, turkey oak, sand post oak, and blackjack oak. The wood is used for fuel and fence posts. Acorns of all oaks rank at the top of wildlife food plants because of their wide distribution, availability, and palatability, and are considered the staff of life for many wildlife species. Acorns of the white oak group are generally preferred over the red oak group because of lower tannic acid levels. Smaller birds and mammals prefer the species producing the smaller acorns, whereas larger birds and mammals will eat all sizes, including the largest. A partial list of acorn eaters include mallard, black, pintail, and wood ducks; northern bobwhite; wild turkey; flying, fox, and gray squirrels; all species of deer; black bear; peccary; raccoon; numerous songbirds; and many small mammals.

Quercus is Latin for "oak tree"; *incana* means "gray" and refers to the grayish pubescence on the leaf underside.

SIMILAR SPECIES Myrtle oak (*Quercus myrtifolia* Willd.) is a shrub or small tree found on the coast of Alabama in pine-oak scrub forests. It has a similar leaf shape to bluejack oak, but the leaves are 2–8 cm (0.8–3.1 in) long with revolute margins and a mostly glabrous underside.

Turkey Oak

Quercus laevis Walter

COMMON NAMES turkey oak, scrub oak

QUICK GUIDE Leaves alternate, simple, with three
to seven bristle-tipped lobes, three-lobed leaves
resemble a turkey foot track, base tapered, petiole twisted and hangs vertically; acorn cap with a
rolled lip; bark dark and rough.

DESCRIPTION Leaves are alternate, simple, deciduous, 8–20 cm (3.1–
7.9 in) long, and often hang vertically, with three to seven bristle-
tipped lobes; lateral lobes are often curved; three-lobed leaves resemble a turkey foot track; sinuses are irregular and may extend more
than halfway to the midrib; base is acute to cuneate; midrib is curved
on some leaves; upperside is shiny; underside has pubescence in vein
axils; autumn color is red. Twigs are stout, red-brown, and glabrous or
pubescent; leaf scar is crescent shaped to oval with numerous bundle
scars. Terminal buds are about 1.2 cm (0.5 in) long and acute; scales
are overlapping and red-brown with rusty pubescence. Bark is dark
gray to black, rough, and possibly blocky. Flowers are imperfect and
appear in the spring with the leaves; staminate flowers are in drooping catkins; pistillate spikes are in leaf axils. Fruit is an acorn, about
2 cm (0.8 in) long; nut is pubescent; cap is bowl-like with appressed,

From left to right:
Turkey oak leaf.
Turkey oak fruit.

From left to right:

Turkey oak bark.

Turkey oak leaves with turkey track outline; photo by Lisa J. Samuelson.

pubescent scales and a rolled edge covering one-third of the nut; fruit matures in two seasons. The growth form is up to 15 m (50 ft) tall but usually smaller.

HABITAT Well-drained, sandy soils.

NOTES Turkey oak can form pure stands or is found in mixed stands with sand hickory, bluejack oak, sand post oak, blackjack oak, sand pine, and longleaf pine. The wood has no commercial value and is used mostly for fuel. Acorns of all oaks rank at the top of wildlife food plants because of their wide distribution, availability, and palatability, and are considered the staff of life for many wildlife species. Acorns of the white oak group are generally preferred over the red oak group because of lower tannic acid levels. Smaller birds and mammals prefer the species producing the smaller acorns, whereas larger birds and mammals will eat all sizes, including the largest. A partial list of acorn eaters include mallard, black, pintail, and wood ducks; northern bobwhite; wild turkey; flying, fox, and gray squirrels; all species of deer; black bear; peccary; raccoon; numerous songbirds; and many small mammals.

Quercus is Latin for "oak tree"; *laevis* means "smooth," perhaps referring to the shiny leaves.

Swamp Laurel Oak

Quercus laurifolia Michx.

COMMON NAMES swamp laurel oak, diamond-leaf oak, laurel oak

QUICK GUIDE Leaves alternate, simple, variable in shape but often spatulate or rhombic, margin entire; bark gray-black and shallowly ridged.

DESCRIPTION Leaves are alternate, simple, persistent, variable in shape (spatulate, obovate, subrhombic, or oblanceolate), 5–14 cm (2.0–5.5 in) long, and leathery; apex may or may not have a bristle tip; base is cuneate; margin is usually entire or occasionally has a coarse tooth or shallow lobe; midrib is yellow; vein axils have tufts of hair. Twigs are moderately stout, gray-brown, and glabrous; leaf scar is crescent shaped to oval with numerous bundle scars. Terminal buds are about 3 mm (0.1 in) long and ovoid; scales are overlapping and red-brown with pubescent margins. Flowers are imperfect and appear in the spring with the leaves; staminate flowers are in drooping catkins; pistillate spikes are in leaf axils. Fruit is an acorn, about 1.5 cm (0.6 in) long; nut is brown, flat topped, and pubescent; cap is flat with appressed pubescent scales covering one-fourth to one-half of the nut; fruit matures in two seasons. Bark is gray-brown to black with flattened sometimes white ridges and becomes blocky on large trees. The growth form is up to 30 m (100 ft) in height, often with a swollen, buttressed trunk.

From left to right:

Swamp laurel oak leaves.

Swamp laurel oak fruit.

From left to right:

Swamp laurel oak bark.

Swamp laurel oak leaves showing subrhombic shape.

HABITAT Low flats, bottomlands, and swamp edges.

NOTES Swamp laurel oak is found growing with red maple, swamp cyrilla, sweetbay magnolia, swamp tupelo, pond pine, overcup oak, swamp chestnut oak, and live oak. It is used as red oak lumber and for pulpwood and fuel. Acorns of all oaks rank at the top of wildlife food plants because of their wide distribution, availability, and palatability, and are considered the staff of life for many wildlife species. Acorns of the white oak group are generally preferred over the red oak group because of lower tannic acid levels. Smaller birds and mammals prefer the species producing the smaller acorns, whereas larger birds and mammals will eat all sizes, including the largest. A partial list of acorn eaters include mallard, black, pintail, and wood ducks; northern bobwhite; wild turkey; flying, fox, and gray squirrels; all species of deer; black bear; peccary; raccoon; numerous songbirds; and many small mammals.

Quercus is Latin for "oak tree"; *laurifolia* means "laurel tree leaves." There is disagreement in the naming of this species, and it is sometimes referred to as *Quercus hemisphaerica* or *Quercus obtusa.*

Overcup Oak

Quercus lyrata Walter

COMMON NAMES overcup oak, swamp post oak, water white oak

QUICK GUIDE Leaves alternate, simple, with five to nine irregularly shaped lobes without bristles; acorn round and almost completely enclosed by a thin cap with fused scales; bark gray-brown with scaly ridges.

DESCRIPTION Leaves are alternate, simple, deciduous, and 12–20 cm (4.7–7.9 in) long, with five to nine bristleless and irregularly shaped lobes; sinuses are irregular in depth and may extend more than half-way to the midrib; base is cuneate; underside is pubescent; autumn color is yellow-brown. Twigs are slender, brown-gray, and glabrous; leaf scar is crescent shaped to oval with numerous bundle scars. Terminal buds are about 3 mm (0.1 in) long and round; scales are overlapping, brown, and pubescent. Flowers are imperfect and appear in the spring with the leaves; staminate flowers are in drooping catkins; pistillate spikes are in leaf axils. Fruit is an acorn, about 2 cm (0.8 in) long, round, and buoyant; nut is shiny brown; cap is loose and thin with fused scales that sometimes split and almost completely cover the nut; fruit matures in one season. Bark is gray-brown and shallowly grooved, with scaly ridges or plates. The growth form is up to 30 m (100 ft) in height and 1 m (3 ft) in diameter.

HABITAT Wet bottomlands, floodplains, and shallow swamps as well as on the edges of deep swamps, rivers, sloughs, and sink holes.

NOTES Overcup oak is tolerant of flooding, and its acorn is adapted for dispersal in water. Forest associates include red maple, water hickory,

From left to right:

Overcup oak leaves

Overcup oak fruit.

Clockwise from left:

Overcup oak bark.

Lobing and sinuses of bur oak leaves for comparison.

Larger bur oak acorn and cap for comparison.

bitternut hickory, shagbark hickory, sugarberry, common persimmon, many ash species, sweetgum, swamp tupelo, swamp laurel oak, cherrybark oak, willow oak, Nutall oak, live oak, and American elm. The wood is used as low-quality white oak lumber. Acorns of all oaks rank at the top of wildlife food plants because of their wide distribution, availability, and palatability, and are considered the staff of life for many wildlife species. Acorns of the white oak group are generally preferred over the red oak group because of lower tannic acid levels. Smaller birds and mammals prefer the species producing the smaller acorns, whereas larger birds and mammals will eat all sizes, including the largest. A partial list of acorn eaters include mallard, black, pintail, and wood ducks; northern bobwhite; wild turkey; flying, fox, and gray squirrels; all species of deer; black bear; peccary; raccoon; numerous songbirds; and many small mammals. Overcup oak is planted as an ornamental tree, but the acorns can be messy.

Quercus is Latin for "oak tree"; *lyrata* means "lyrelike."

SIMILAR SPECIES **Bur oak** (*Quercus macrocarpa* Michx.) leaves are somewhat similar to overcup oak and have five to nine bristleless lobes, but the terminal lobe is fan shaped, and on some leaves, the middle sinus is the deepest and elongated. The twigs have corky wings and the acorn is the largest of the native oaks (2–5 cm [0.8–2.0 in] long), with a fringed cap that covers from half to most of the downy nut. Bur oak is uncommon in Alabama but is planted in urban landscapes due to its tolerance of drought.

Sand Post Oak

Quercus margarettiae Ashe ex Small

COMMON NAMES sand post oak, scrubby post oak

QUICK GUIDE Leaves alternate, simple, leathery, with three to five bristleless lobes pointing toward the apex, underside pubescent; twigs mostly glabrous; bark gray-brown with scaly ridges.

DESCRIPTION Leaves are alternate, simple, deciduous, 2–15 cm (0.8–5.9 in) long, and leathery, with three to five lobes that lack bristles and usually point toward the apex; central lobes are only occasionally crosslike; some leaves may be unlobed; sinuses are shallow and variable; base is rounded to cuneate; underside is pubescent; autumn color is yellow-brown or dull red. Twigs are slender, red-brown, and glabrous or with some pubescence; leaf scar is crescent shaped to oval with numerous bundle scars. Terminal buds are about 3 mm (0.1 in) long, ovoid, and angled; scales are overlapping, red-brown, and glabrous or with some pubescence. Flowers are imperfect and appear in the spring with the leaves; staminate flowers are in drooping catkins; pistillate spikes are in leaf axils. Fruit is an acorn, 1.3–2.0 cm (0.5–0.8 in) long; nut is pubescent; cap is bowl shaped with pubescent, appressed scales covering one-third to one-half of the nut; fruit matures in one season. Bark is gray-brown with scaly ridges. The growth form is up to 18 m (60 ft) in height but usually smaller.

From left to right:

Sand post oak leaves.

Sand post oak fruit.

From left to right:

Sand post oak bark.

Sand post oak leaves showing variation in lobing; photo by Lisa J. Samuelson.

HABITAT Well-drained uplands and dry sandy soils.

NOTES On drier sites, sand post oak is associated with sand hickory, sand pine, longleaf pine, bluejack oak, turkey oak, and blackjack oak. The wood is not commercially important. Acorns of all oaks rank at the top of wildlife food plants because of their wide distribution, availability, and palatability, and are considered the staff of life for many wildlife species. Acorns of the white oak group are generally preferred over the red oak group because of lower tannic acid levels. Smaller birds and mammals prefer the species producing the smaller acorns, whereas larger birds and mammals will eat all sizes, including the largest. A partial list of acorn eaters include mallard, black, pintail, and wood ducks; northern bobwhite; wild turkey; flying, fox, and gray squirrels; all species of deer; black bear; peccary; raccoon; numerous songbirds; and many small mammals.

Quercus is Latin for "oak tree"; *margarettiae* is for Margaret Ashe.

Blackjack Oak

Quercus marilandica Muenchh.

COMMON NAMES blackjack oak, black oak, jack oak

QUICK GUIDE Leaves alternate, simple, fan shaped, leathery, apex with three broad and bristle-tipped lobes; acorn cap densely hairy and covering up to one-half of the nut; bark dark, rough, and blocky.

DESCRIPTION Leaves are alternate, simple, deciduous, T-shaped or fan shaped, 8–20 cm (3.1–7.9 in) long, and very leathery; apex has three broad and bristle-tipped lobes; base is acute to cordate; underside has orange or yellow pubescence; autumn color is yellow-brown. Twigs are stout, brown-gray, and pubescent; leaf scar is crescent shaped to oval with numerous bundle scars. Terminal buds are about 6 mm (0.2 in) long, acute, and angled; scales are overlapping, red-brown, and pubescent. Flowers are imperfect and appear in the spring after the leaves; staminate flowers are in drooping catkins; pistillate spikes are in leaf axils. Fruit is an acorn, about 2 cm (0.8 in) long; nut is pubescent with a rigid point; cap is bowl shaped with shaggy pubescent scales covering one-third to one-half of the nut; fruit matures in two seasons. Bark is gray-black and rough or blocky. The growth form is up to 15 m (50 ft) in height.

HABITAT Sandhills and dry upland forests.

From left to right:

Blackjack oak leaves; photo by Lisa J. Samuelson.

Blackjack oak fruit; photo by Lisa J. Samuelson.

Blackjack oak bark; photo by Lisa J. Samuelson.

NOTES On dry sandy soils, blackjack oak is found growing with sand hickory, longleaf pine, bluejack oak, turkey oak, and sand post oak. On upland soils, it is associated with sourwood, southern red oak, chestnut oak, and shortleaf pine. The wood is used as red oak lumber and for fuel, fences, and railroad ties. Acorns of all oaks rank at the top of wildlife food plants because of their wide distribution, availability, and palatability, and are considered the staff of life for many wildlife species. Acorns of the white oak group are generally preferred over the red oak group because of lower tannic acid levels. Smaller birds and mammals prefer the species producing the smaller acorns, whereas larger birds and mammals will eat all sizes, including the largest. A partial list of acorn eaters include mallard, black, pintail, and wood ducks; northern bobwhite; wild turkey; flying, fox, and gray squirrels; all species of deer; black bear; peccary; raccoon; numerous songbirds; and many small mammals.

Quercus is Latin for "oak tree"; *marilandica* refers to the geographic range.

SIMILAR SPECIES **Arkansas oak** (*Quercus arkansana* Sarg.) is a smaller understory tree up to 15 m (49 ft) tall found in patches around creek heads mostly in central Alabama. It is distinguished by leaves similar to blackjack oak but smaller, thinner, and more broadly three-lobed or unlobed at the apex, and a glabrous or pubescent underside. Scales on the acorn cap are thinner and smaller than in blackjack oak and cover about one-third of the nut.

Swamp Chestnut Oak

Quercus michauxii Nutt.

COMMON NAMES swamp chestnut oak, cow oak, basket oak

QUICK GUIDE Leaves alternate, simple, obovate throughout the canopy, margin scalloped with round and callous-tipped teeth; acorn large with a scaly cap; bark gray-brown and scaly.

DESCRIPTION Leaves are alternate, simple, deciduous, obovate, and 12–22 cm (4.7–8.7 in) long; apex may be abruptly acuminate; base is cuneate to rounded; margin is scalloped with rounded and callous-tipped teeth; underside is pubescent; autumn color is yellow-brown to dull red. Twigs are stout, red-brown to gray-brown, and glabrous to densely pubescent; leaf scar is crescent shaped to oval with numerous bundle scars. Terminal buds are up to 6 mm (0.2 in) long and ovoid; scales are overlapping, red-brown, and lightly pubescent. Flowers are imperfect and appear in the spring with the leaves; staminate flowers are in drooping catkins; pistillate spikes are in leaf axils. Fruit is an acorn, 2.5–4.0 cm (1.0–1.6 in) long; cap is bowl shaped with pubescent scales and a thin edge covering one-third of the nut; fruit matures in one season. Bark is light gray to gray-brown and scaly. The growth form is up to 30 m (100 ft) in height with a long, straight bole.

From left to right:

Swamp chestnut oak leaves.

Swamp chestnut oak fruit.

Swamp chestnut oak bark.

HABITAT Well-drained bottomlands and along the edges of streams and swamps.

NOTES Swamp chestnut oak is found with red maple, many ashes, sweetgum, bitternut hickory, shagbark hickory, spruce pine, and bottomland oaks. The wood is used as white oak lumber. Thin strips are cut from saplings and used in making baskets. Acorns of all oaks rank at the top of wildlife food plants because of their wide distribution, availability, and palatability, and are considered the staff of life for many wildlife species. Acorns of the white oak group are generally preferred over the red oak group because of lower tannic acid levels. Smaller birds and mammals prefer the species producing the smaller acorns, whereas larger birds and mammals will eat all sizes, including the largest. A partial list of acorn eaters include mallard, black, pintail, and wood ducks; northern bobwhite; wild turkey; flying, fox, and gray squirrels; all species of deer; black bear; peccary; raccoon; numerous songbirds; and many small mammals.

Quercus is Latin for "oak tree"; *michauxii* is for the French botanist F. Michaux.

Chestnut Oak

Quercus montana Willd.

COMMON NAMES chestnut oak, tanbark oak, rock oak

QUICK GUIDE Leaves alternate, simple, obovate, margin scalloped with rounded smooth teeth; acorn cap thin with warty scales; bark dark gray and deeply grooved with orange in the grooves.

DESCRIPTION Leaves are alternate, simple, deciduous, obovate to elliptical, and 10–22 cm (3.9–8.7 in) long; base is acute; margin is scalloped with rounded teeth with no bristles or callous tips; underside is pale and pubescent or glabrous; autumn color is yellow-brown to dull red. Twigs are stout, red-brown, and glabrous; leaf scar is crescent shaped to oval with numerous bundle scars. Terminal buds are 6–13 mm (0.2–1.5 in) long and acute; scales are overlapping, chestnut brown, and lightly pubescent on the margins. Flowers are imperfect and appear in the spring with the leaves; staminate flowers are in drooping catkins; pistillate spikes are in leaf axils. Fruit is an acorn, oblong, and 2–4 cm (0.8–1.6 in) long; nut is shiny and yellow-brown when young; cap is bowl shaped with fused, warty scales and a thin edge covering one-third of the nut; cap becomes loose on the nut when the nut matures; fruit matures in one season. Bark is gray and smooth or shallowly grooved on small trees; bark on large trees is deeply furrowed with orange or orange-brown in the grooves. The growth form is up to 24 m (80 ft) in height and 1 m (3 ft) in diameter.

From left to right:

Chestnut oak leaves.

Chestnut oak fruit.

From left to right:

Chestnut oak bark.

Chestnut oak bark with orange in the grooves.

HABITAT Sandy or rocky uplands.

NOTES Forest associates of chestnut oak include red maple, upland hickories, eastern redcedar, sweetgum, blackgum, sourwood, short-leaf pine, Virginia pine, and other upland oaks such as scarlet oak, chinkapin oak, and post oak. Chestnut oak is used as white oak lumber. The thick bark was used as a source of tannin. Acorns of all oaks rank at the top of wildlife food plants because of their wide distribution, availability, and palatability, and are considered the staff of life for many wildlife species. Acorns of the white oak group are generally preferred over the red oak group because of lower tannic acid levels. Smaller birds and mammals prefer the species producing the smaller acorns, whereas larger birds and mammals will eat all sizes, including the largest. A partial list of acorn eaters include mallard, black, pintail, and wood ducks; northern bobwhite; wild turkey; flying, fox, and gray squirrels; all species of deer; black bear; peccary; raccoon; numerous songbirds; and many small mammals. The sweet acorn is preferred by wildlife.

Quercus is Latin for "oak tree"; *montana* means "mountain." This species is also referred to as *Quercus prinus* L. but see Whittemore and Nixon (2005).

Chinkapin Oak

Quercus muehlenbergii Engelm.

COMMON NAMES chinkapin oak, scrub chestnut oak, yellow oak

QUICK GUIDE Leaves alternate, simple, elliptical, margin scalloped with curved and callous-tipped teeth; acorn cap with appressed scales covering one-third of the nut; bark gray to gray-brown and scaly.

DESCRIPTION Leaves are alternate, simple, deciduous, obovate in the lower canopy, more elliptical or lanceolate in the upper canopy, and 9–18 cm (3.5–7.1 in) long; apex is often acuminate; base is acute; margin is scalloped with curved, callous-tipped teeth, but sometimes the margin is only wavy; underside is pubescent; autumn color is yellow-brown or dull red. Twigs are slender, red-brown to gray, and glabrous or lightly pubescent; leaf scar is crescent shaped to oval with numerous bundle scars. Terminal buds are 3–5 mm (0.1–0.2 in) long and acute; scales are overlapping and red-brown, with white pubescence on the

Clockwise from upper left:

Leaves of chinkapin oak.

Chinkapin oak fruit.

Obovate chinkapin oak leaf.

Chinkapin oak leaf margin with callous-tipped teeth; photo by Lisa J. Samuelson.

Chinkapin oak bark.

margins, making them appear striped. Flowers are imperfect and appear in the spring with the leaves; staminate flowers are in drooping catkins; pistillate spikes are in leaf axils. Fruit is an acorn, 1.3–2.5 cm (0.6–1.0 in) long; nut is brown to black; cap is thin and bowl shaped with appressed, pubescent scales and a thin edge covering one-third to one-half of the nut; fruit matures in one season. Bark is gray to gray-brown and scaly. The growth form is up to 24 m (80 ft) in height and 1 m (3 ft) in diameter.

HABITAT Alkaline soils, bluffs, upland slopes, and stream edges.

NOTES Forest associates of chinkapin oak include red maple, many hickories, American beech, eastern redcedar, tulip-poplar, shortleaf pine, Virginia pine, white oak, scarlet oak, southern red oak, chestnut oak, northern red oak, black oak, and winged elm. The wood is used as white oak lumber and for construction and fuel. Acorns of all oaks rank at the top of wildlife food plants because of their wide distribution, availability, and palatability, and are considered the staff of life for many wildlife species. Acorns of the white oak group are generally preferred over the red oak group because of lower tannic acid levels. Smaller birds and mammals prefer the species producing the smaller acorns, whereas larger birds and mammals will eat all sizes, including the largest. A partial list of acorn eaters include mallard, black, pintail, and wood ducks; northern bobwhite; wild turkey; flying, fox, and gray squirrels; all species of deer; black bear; peccary; raccoon; numerous songbirds; and many small mammals.

Quercus is Latin for "oak tree"; *muehlenbergii* is for the American botanist Muehlenberg.

Water Oak

Quercus nigra L.

COMMON NAMES water oak, opossum oak, fiddle oak

QUICK GUIDE Leaves alternate, simple, spatulate, margin usually entire, apex with a bristle tip, underside mostly glabrous; leaves of seedlings and saplings variable in shape and often pronged; acorn cap covering the base of the nut; bark brown-gray and mostly smooth or shallowly ridged.

DESCRIPTION Leaves are alternate, simple, persistent, obovate to spatulate but variable, 5–10 cm (2.0–3.9 in) long, leathery, and nearly sessile; apex is obtuse to rounded and bristle tipped; base is acute to cuneate; margin is usually entire or may be lobed near the apex; underside is mostly glabrous; autumn color is yellow-brown. Leaves of seedlings, saplings, and root sprouts are variable in shape and often larger, more oblong, and lobed. Twigs are slender, red-brown, and glabrous; leaf scar is crescent shaped to oval with numerous bundle scars. Terminal buds are about 5 mm (0.2 in) long and acute; scales are overlapping, gray-brown to red-brown, and glabrous or pubescent.

Clockwise from upper left:

Water oak leaves.

Leaves of a sapling water oak; photo by Lisa J. Samuelson.

Water oak pistillate flower.

Water oak fruit.

Bark of a large water oak tree.

Flowers are imperfect and appear in the spring with the leaves; staminate flowers are in drooping catkins; pistillate spikes are in leaf axils. Fruit is an acorn, about 1.2 cm (0.5 in) long; nut is dark brown, pubescent, and flat topped; cap is flat with appressed, pubescent scales covering one-fourth of the nut; fruit matures in two seasons. Bark is brown-gray and smooth or only shallowly ridged. The growth form is up to 36 m (120 ft) in height, possibly with mistletoe in the canopy.

HABITAT Ubiquitous, a variety of sites from dry uplands to moderately wet bottomlands.

NOTES Water oak is a common oak found growing with a great variety of forest associates depending on the site. The wood is used as red oak lumber and for construction, boxes, and fuel. Acorns of all oaks rank at the top of wildlife food plants because of their wide distribution, availability, and palatability, and are considered the staff of life for many wildlife species. Acorns of the white oak group are generally preferred over the red oak group because of lower tannic acid levels. Smaller birds and mammals prefer the species producing the smaller acorns, whereas larger birds and mammals will eat all sizes, including the largest. A partial list of acorn eaters include mallard, black, pintail, and wood ducks; northern bobwhite; wild turkey; flying, fox, and gray squirrels; all species of deer; black bear; peccary; raccoon; numerous songbirds; and many small mammals.

Quercus is Latin for "oak tree"; *nigra* means "dark," referring to the nut.

Cherrybark Oak

Quercus pagoda Raf.

COMMON NAMES cherrybark oak, red oak, swamp red oak, bottomland red oak

QUICK GUIDE Leaves alternate, simple, with five to seven bristle-tipped lobes, basal lobes prominent, base tapered, underside densely hairy; acorn cap covering up to one-half of the pubescent nut; bark gray-brown and flaky or with scaly ridges.

DESCRIPTION Leaves are alternate, simple, deciduous, and 13–23 cm (5.1–9.1 in) long, with five to seven bristle-tipped lobes; basal lobes are prominent and give the leaf a pagoda shape; base is truncate to tapered; underside is densely pubescent; autumn color is yellow-brown. Sinuses are shallow on shade leaves and deeper on sun leaves. Sun leaves are similar to southern red oak except the base is not bell shaped. Twigs are green to red-brown with gray pubescence; leaf scar is crescent shaped to oval with numerous bundle scars. Terminal buds are about 5 mm (0.2 in) long, ovoid, and somewhat angled; scales are overlapping, red-brown, and pubescent. Flowers are imperfect and appear in the spring with the leaves; staminate flowers are in drooping catkins; pistillate spikes are in leaf axils. Fruit is an acorn, about 1.3 cm (0.5 in) long; nut is pubescent; cap is bowl-like with appressed, pubescent scales covering one-third to one-half of the nut; fruit matures in two seasons. Bark on small trees is smooth and on larger trees is gray-brown and flaky,

From left to right:

Sun leaves of cherrybark oak.

Shade leaves of cherrybark oak.

Cherrybark oak fruit.

From left to right:

Flaky bark of cherrybark oak .

Furrowed bark of cherrybark oak.

somewhat like black cherry. On very large trees the bark is shallowly furrowed with scaly ridges. The growth form is up to 39 m (130 ft) in height and 1.5 m (5 ft) in diameter with a long straight bole.

HABITAT Bottomlands, floodplain forests, terraces, stream margins, and swamp edges.

NOTES Cherrybark oak is found with bitternut hickory, shagbark hickory, numerous ashes, sweetgum, spruce pine, overcup oak, swamp chestnut oak, water oak, willow oak, Shumard oak, black walnut, and American elm. The red oak wood is commercially important and used for trim, finish work, flooring, and furniture. Acorns of all oaks rank at the top of wildlife food plants because of their wide distribution, availability, and palatability, and are considered the staff of life for many wildlife species. Acorns of the white oak group are generally preferred over the red oak group because of lower tannic acid levels. Smaller birds and mammals prefer the species producing the smaller acorns, whereas larger birds and mammals will eat all sizes, including the largest. A partial list of acorn eaters include mallard, black, pintail, and wood ducks; northern bobwhite; wild turkey; flying, fox, and gray squirrels; all species of deer; black bear; peccary; raccoon; numerous songbirds; and many small mammals.

Quercus is Latin for "oak tree"; *pagoda* refers to the pagoda-shaped leaves.

Willow Oak

Quercus phellos L.

COMMON NAMES willow oak, swamp willow oak, peach oak

QUICK GUIDE Leaves alternate, simple, lanceolate, thin, often drooping, apex with a bristle tip, margin entire, midrib often with tufts of hair; acorn cap covering the base of the nut; bark gray and smooth or shallowly ridged.

DESCRIPTION Leaves are alternate, simple, deciduous, linear to lanceolate and "willowlike," 5–12 cm (2.0–4.7 in) long, thin, and nearly sessile; apex is acute with a bristle tip; base is acute; margin is entire or sometimes wavy; midrib is yellow and often has tufts of hair; autumn color is yellow-brown. Twigs are slender, gray-brown, and glabrous; leaf scar is crescent shaped to oval with numerous bundle scars. Terminal buds are about 3 mm (0.1 in) long and ovoid to acute; scales are overlapping, red-brown, and glabrous. Flowers are imperfect and appear in the spring with the leaves; staminate flowers are in drooping catkins; pistillate spikes are in leaf axils. Fruit is an acorn, 6–13 mm (0.2–0.5 in) long; nut is flat topped; cap has appressed scales covering one-fourth or less of the nut; fruit matures in two seasons. Bark is brown-gray and smooth and becomes shallowly ridged; bark of very large trees is scaly

Clockwise from upper left:

Willow oak leaves.

Willow oak staminate flowers.

Willow oak fruit.

From left to right:

Willow oak bark.

Bark of large willow oak tree.

or has blocky ridges. The growth form is up to 36 m (120 ft) in height and 1 m (3 ft) in diameter. Willow oak can sprout from detached roots.

HABITAT Moist to poorly drained soils of bottomlands and along stream and swamp edges.

NOTES Willow oak can be found growing with red maple, sugarberry, common persimmon, green ash, sweetgum, eastern cottonwood, boxelder, bottomland oaks, and American elm. Willow oak is planted to control soil erosion along water margins. The wood is used as red oak lumber and for pulpwood and fuel. Willow oak is a popular ornamental tree. Acorns of all oaks rank at the top of wildlife food plants because of their wide distribution, availability, and palatability, and are considered the staff of life for many wildlife species. Acorns of the white oak group are generally preferred over the red oak group because of lower tannic acid levels. Smaller birds and mammals prefer the species producing the smaller acorns, whereas larger birds and mammals will eat all sizes, including the largest. A partial list of acorn eaters include mallard, black, pintail, and wood ducks; northern bobwhite; wild turkey; flying, fox, and gray squirrels; all species of deer; black bear; peccary; raccoon; numerous songbirds; and many small mammals.

Quercus is Latin for "oak tree"; *phellos* means "cork tree."

Northern Red Oak

Quercus rubra L.

COMMON NAMES northern red oak, eastern red oak, red oak

QUICK GUIDE Leaves alternate, simple, with 7 to 11 bristle-tipped lobes, sinuses shallow; buds mostly glabrous and chestnut brown; acorn cap beanie-like; bark gray-black with flat often white ridges.

DESCRIPTION Leaves are alternate, simple, deciduous, and 10–23 cm (3.9–9.1 in) long, with 7 to 11 bristle-tipped lobes; sinuses are shallow and extend at most halfway to the midrib; base is truncate to acute; underside is mostly glabrous with tufts of hair in vein axils; autumn color is red. Twigs are red-brown and glabrous; leaf scar is crescent shaped to oval with numerous bundle scars. Terminal buds are about 7 mm (0.3 in) long, acute, and slightly angled; scales are overlapping and chestnut brown with minute pubescence on the margins. Flowers are imperfect and appear in the spring with the leaves; staminate flowers are in drooping catkins; pistillate spikes are in leaf axils. Fruit is an acorn, 2–3 cm (0.8–1.2 in) long; nut is pubescent; cap has appressed and pubescent or glabrous scales and is beanie-like with rolled edges, covering at most one-fourth of the nut; fruit matures in two seasons. Bark is gray-black and shallowly grooved or has flattened gray or white ridges like ski tracks; bark of large trees is more furrowed; inner bark is orange-brown. The growth form is up to 30 m (100 ft) in height and 1 m (3 ft) in diameter with a straight bole.

HABITAT Moist, well-drained soils.

From left to right:

Northern red oak leaves.

Northern red oak fruit.

From left to right:

Northern red oak bark with white "ski tracks."

Bark of large northern red oak.

NOTES Northern red oak can be found with many species, including red maple, sugar maple, yellow buckeye, hickories, American beech, green and white ash, tulip-poplar, blackgum, eastern white pine, loblolly pine, black cherry, white oak, black oak, basswood, eastern hemlock, and American elm. The "red oak wood" is red-brown, heavy, hard, and commercially important, and is used for trim, flooring, cabinets, furniture, and caskets. Northern red oak is planted as an ornamental tree for its fall color and site adaptability. Acorns of all oaks rank at the top of wildlife food plants because of their wide distribution, availability, and palatability, and are considered the staff of life for many wildlife species. Acorns of the white oak group are generally preferred over the red oak group because of lower tannic acid levels. Smaller birds and mammals prefer the species producing the smaller acorns, whereas larger birds and mammals will eat all sizes, including the largest. A partial list of acorn eaters include mallard, black, pintail, and wood ducks; northern bobwhite; wild turkey; flying, fox, and gray squirrels; all species of deer; black bear; peccary; raccoon; numerous songbirds; and many small mammals.

Quercus is Latin for "oak tree"; *rubra* means "red."

Shumard Oak

Quercus shumardii Buckley

COMMON NAMES Shumard oak, red oak, swamp red oak

QUICK GUIDE Leaves alternate, simple, with five to nine lobes with many bristle tips, sinuses deep; acorn cap covering the top of the striped nut; bark gray-brown and smooth or shallowly fissured.

DESCRIPTION Leaves are alternate, simple, deciduous, and 10–20 cm (3.9–7.9 in) long, with five to nine lobes and many bristle tips; sinuses are U-shaped and extend halfway or more to the midrib; base is truncate to obtuse; vein axils have tufts of hair; autumn color is red. Twigs are brown to gray-brown and glabrous; leaf scar is crescent shaped to oval with numerous bundle scars. Terminal buds are about 6 mm (0.2 in) long, acute, and angled; scales are overlapping and gray-brown with gray pubescence. Flowers are imperfect and appear in the spring with the leaves; staminate flowers are in drooping catkins; pistillate spikes are in leaf axils. Fruit is an acorn, 2–3 cm (0.8–1.2 in) long; nut is striped and pubescent; cap has appressed pubescent or glabrous scales and rolled edges and covers up to one-third of the nut; fruit matures in two seasons. Bark is gray-brown and smooth on small trees; bark of large trees is shallowly fissured and may have scaly ridges. The growth form is up to 30 m (100 ft) in height and 1.2 m (4 ft) in diameter.

From left to right:

Shumard oak leaf.

Shumard oak fruit.

Shumard oak bark.

HABITAT Next to streams, rivers, and swamps; in bottomlands; and on limestone soils.

NOTES Shumard oak is found in association with water hickory, shagbark hickory, bitternut hickory, American beech, numerous ashes, spruce pine, bottomland oaks, and American elm. The red oak wood is commercially important and used for trim, flooring, cabinets, and furniture. Acorns of all oaks rank at the top of wildlife food plants because of their wide distribution, availability, and palatability, and are considered the staff of life for many wildlife species. Acorns of the white oak group are generally preferred over the red oak group because of lower tannic acid levels. Smaller birds and mammals prefer the species producing the smaller acorns, whereas larger birds and mammals will eat all sizes, including the largest. A partial list of acorn eaters include mallard, black, pintail, and wood ducks; northern bobwhite; wild turkey; flying, fox, and gray squirrels; all species of deer; black bear; peccary; raccoon; numerous songbirds; and many small mammals.

Quercus is Latin for "oak tree"; *shumardii* is for the geologist Shumard.

Post Oak

Quercus stellata Wangenh.

COMMON NAMES post oak, iron oak

QUICK GUIDE Leaves alternate, simple, leathery, with five bristleless lobes, central lobes may form a cross, pubescent upper surface and underside; twigs pubescent; acorn nut and cap downy pubescent; bark gray-brown with scaly ridges.

DESCRIPTION Leaves are alternate, simple, deciduous, 10–15 cm (3.9–5.9 in) long, and leathery, with five bristleless lobes; lobes are variable in shape and central lobes may be squarish and form a cross; sinuses are variable in depth and length; base is cuneate; upper surface has stellate pubescence; underside is densely pubescent to tomentose; autumn color is yellow-brown. Twigs are stout, red-brown to gray, and densely pubescent; leaf scar is crescent shaped to oval with numerous bundle scars. Terminal buds are ovoid, 3 mm (0.1 in) long, and angled; scales are overlapping, red-brown, and glabrous or with yellow-brown pubescence. Flowers are imperfect and appear in the spring with the leaves; staminate flowers are in drooping catkins; pistillate spikes are in leaf axils. Fruit is an acorn, 1.3–2.0 cm (0.5–0.8 in) long; nut is pubescent and dark; cap is bowl shaped and pubescent, with loosely appressed scales covering one-third to one-half of the nut; fruit

Clockwise from left:

Post oak leaf with cruciform central lobes.

Post oak fruit.

Post oak leaves with variable lobing.

Bark of large
post oak.

matures in one season. Bark is gray-brown with scaly ridges on small
trees; bark of large trees has more thickened and irregular ridges
and can be deeply grooved. The growth form is up to 30 m (100 ft) in
height and 1.2 m (4 ft) in diameter.

HABITAT A variety of dry sites, common on well-drained rocky and
sandy soils.

NOTES Forest associates of post oak include numerous hickories, eastern redcedar, sweetgum, blackgum, sourwood, shortleaf pine, loblolly pine, Virginia pine, and other upland oaks. Post oak is used as
white oak lumber and for construction, pulpwood, veneer, railroad ties,
posts, and fuel. Acorns of all oaks rank at the top of wildlife food plants
because of their wide distribution, availability, and palatability, and
are considered the staff of life for many wildlife species. Acorns of the
white oak group are generally preferred over the red oak group because
of lower tannic acid levels. Smaller birds and mammals prefer the species producing the smaller acorns, whereas larger birds and mammals
will eat all sizes, including the largest. A partial list of acorn eaters include mallard, black, pintail, and wood ducks; northern bobwhite; wild
turkey; flying, fox, and gray squirrels; all species of deer; black bear;
peccary; raccoon; numerous songbirds; and many small mammals.

Quercus is Latin for "oak tree"; stellata means "star shaped," referring to the leaf hairs.

Nuttall Oak

Quercus texana Buckley

COMMON NAMES Nuttall oak, Texas red oak, striped oak

QUICK GUIDE Leaves alternate, simple, with five to seven bristle-tipped lobes, apical lobe elongated, sinuses deep and variable on the same leaf; acorn nut maroon-brown striped; bark gray-brown and smooth or shallowly grooved.

DESCRIPTION Leaves are alternate, simple, deciduous, and 10–18 cm (3.9–7.1 in) long, with five to seven bristle-tipped lobes; parallel lobes are often uneven; basal lobes are perpendicular to the midrib, but other lobes point toward the apex; apical lobe is often elongated; sinuses are variable on the same leaf and extend more than halfway to the midrib; base is truncate to acute or inequilateral; vein axils have tufts of hair; autumn color is red. Twigs are red-brown to gray-brown and glabrous; leaf scar is crescent shaped to oval with numerous bundle scars. Terminal buds are about 6 mm (0.2 in) long, acute, and slightly angled; scales are overlapping, light gray-brown, and pubescent or glabrous. Flowers are imperfect and appear in the spring with the leaves; staminate flowers are in drooping catkins; pistillate spikes are in leaf axils. Fruit is an acorn, about 3 cm (1.2 in) long; nut is dark, striped, and pubescent; cap has pubescent, appressed scales that cover one-third to one-half of the nut; fruit matures in two seasons. Bark

From left to right:
Nuttall oak leaf.
Nuttall oak fruit.

Nuttall oak bark

is gray-brown and smooth on small trees and shallowly grooved or ridged on large trees. The growth form is up to 36 m (120 ft) in height and 1 m (3 ft) in diameter.

HABITAT Bottomlands and wet clay flats.

NOTES Nutall oak is an occasional tree in Alabama found in wet woods with red maple, green ash, sweetgum, blackgum, swamp tupelo, overcup oak, swamp chestnut oak, water oak, willow oak, and American elm. The wood is used as red oak lumber and for construction, railroad ties, and fuel. Nutall oak is a popular shade tree and planted in urban areas. Acorns of all oaks rank at the top of wildlife food plants because of their wide distribution, availability, and palatability, and are considered the staff of life for many wildlife species. Acorns of the white oak group are generally preferred over the red oak group because of lower tannic acid levels. Smaller birds and mammals prefer the species producing the smaller acorns, whereas larger birds and mammals will eat all sizes, including the largest. A partial list of acorn eaters include mallard, black, pintail, and wood ducks; northern bobwhite; wild turkey; flying, fox, and gray squirrels; all species of deer; black bear; peccary; raccoon; numerous songbirds; and many small mammals.

Quercus is Latin for "oak tree"; *texana* refers to the geographic range.

Black Oak

Quercus velutina Lam.

COMMON NAMES black oak, yellow oak, yellowbark oak, quercitron oak

QUICK GUIDE Leaves alternate, simple, variable in shape, with five to seven bristle-tipped lobes; buds angled with tawny pubescence; acorn cap slightly fringed and pubescent, covering one-third of the nut; bark dark and ridged; inner bark yellow-orange.

DESCRIPTION Leaves are alternate, simple, deciduous, and 10–23 cm (3.9–9.1 in) long, with five to seven bristle-tipped lobes; base is truncate to obtuse; autumn color is yellow to dull red. Shade leaves have more irregular and shallow sinuses and dense to sparse pubescence on the underside. Sun leaves usually have sinuses reaching halfway or more to the midrib and less pubescence. Twigs are red-brown and glabrous or pubescent; leaf scar is crescent shaped to oval with numerous bundle scars. Terminal buds are 1.3 cm (0.5 in) long, acute, and angled; scales are overlapping and covered with tawny pubescence. Flowers are imperfect and appear in the spring with the leaves; staminate flowers are in drooping catkins; pistillate spikes are in leaf axils. Fruit is an acorn, 1–2 cm (0.4–0.8 in) long; nut is pubescent; cap is pubescent with loosely appressed scales that cover one-third to one-half of the nut; fruit matures in two seasons. Bark is gray-black and rough on small trees; bark on large trees is blockier and more ridged, and the upper trunk may show white flattened ridges; inner bark is

From left to right:

Black oak shade leaves.

Black oak sun leaf.

Black oak fruit.

Black oak bark.

yellow-orange. The growth form is up to 30 m (100 ft) in height and 1.2 m (4 ft) in diameter.

HABITAT Upland sites.

NOTES Forest associates of black oak are numerous and include many upland hickories, oaks, and pines. The red oak wood is commercially important and is used for trim, flooring, cabinets, and furniture. Black oak was once used as a source of tannin. A yellow dye made from the inner bark was used in printing calicoes. Acorns of all oaks rank at the top of wildlife food plants because of their wide distribution, availability, and palatability, and are considered the staff of life for many wildlife species. Acorns of the white oak group are generally preferred over the red oak group because of lower tannic acid levels. Smaller birds and mammals prefer the species producing the smaller acorns, whereas larger birds and mammals will eat all sizes, including the largest. A partial list of acorn eaters include mallard, black, pintail, and wood ducks; northern bobwhite; wild turkey; flying, fox, and gray squirrels; all species of deer; black bear; peccary; raccoon; numerous songbirds; and many small mammals.

Quercus is Latin for "oak tree"; *velutina* means "velvety," perhaps referring to the underside of developing leaves.

Live Oak

Quercus virginiana Mill.

COMMON NAMES live oak, Virginia live oak

QUICK GUIDE Leaves alternate, simple, evergreen, oblong, margin with an occasional bristle-tipped tooth; acorn cap long-stalked; bark gray-brown, deeply grooved, and blocky.

DESCRIPTION Leaves are alternate, simple, evergreen, oblanceolate to oblong or elliptical, 4–13 cm (1.6–5.1 in) long, shiny, leathery, and nearly sessile; apex is obtuse to acute, with or without a rigid bristle tip; base is acute to cuneate; margin is entire and possibly revolute and may show an occasional rigid bristle-tipped tooth; petiole and underside have pale or yellow-brown pubescence. Twigs are slender, gray-brown, and glabrous or pubescent; leaf scar is crescent shaped to oval with numerous bundle scars. Terminal buds are about 2 mm (0.1 in) long and round; scales are overlapping, brown, and lightly pubescent. Flowers are imperfect and appear in the spring with the new leaves; staminate flowers are in short drooping catkins; pistillate spikes are in leaf axils. Fruit is an acorn, 1.0–2.5 cm (0.4–1.0 in) long; nut is shiny dark brown to black; cap is long-stalked and turbinate, with appressed knobby scales that cover one-third to one-half of the nut; fruit matures in one season. Bark is red-brown to brown-gray and deeply fissured with corky blocky ridges. Open-grown trees have a short and heavy trunk. The crown of very old trees may span over 40 m (130 ft) and branches may rest on the ground. The growth form is up to 23 m (75 ft) in height with a spreading crown.

Clockwise from left:

Live oak leaves, some with margin teeth.

Live oak fruit.

Live oak leaves.

From left to right:

Live oak bark.

Live oak form; courtesy of Nancy Loewenstein.

HABITAT Sandy coastal soils ranging from dry to wet.

NOTES Live oak can withstand brief inundation and salt spray. It is found growing with red maple, Atlantic white-cedar, sweetbay magnolia, slash pine, spruce pine, longleaf pine, swamp laurel oak, water oak, baldcypress, and pondcypress. On drier sites it is found growing with sand live oak and scrub oaks. The wood is yellow-brown to white, hard, and heavy, and was used in shipbuilding before the steel era. Live oak is a popular ornamental because of its evergreen leaves, low crown, and the Spanish moss draping the limbs, but it is not cold hardy. Acorns of all oaks rank at the top of wildlife food plants because of their wide distribution, availability, and palatability, and are considered the staff of life for many wildlife species. Acorns of the white oak group are generally preferred over the red oak group because of lower tannic acid levels. Smaller birds and mammals prefer the species producing the smaller acorns, whereas larger birds and mammals will eat all sizes, including the largest. A partial list of acorn eaters include mallard, black, pintail, and wood ducks; northern bobwhite; wild turkey; flying, fox, and gray squirrels; all species of deer; black bear; peccary; raccoon; numerous songbirds; and many small mammals.

Quercus is Latin for "oak tree"; *virginiana* refers to the geographic range.

SIMILAR SPECIES **Sand live oak** (*Quercus geminata* Small) is similar to live oak. It is distinguished by a leaf with a more revolute margin and an underside with dense gray pubescence and by acorns that are short-stalked and often in pairs. Sand live oak is found on deep, sandy soils in the southern portion of Alabama.

Walnut Family (*Juglandaceae*)

Water Hickory

Carya aquatica (Michx. f.) Elliott

COMMON NAMES water hickory, pecan hickory, bitter pecan

QUICK GUIDE Leaves alternate, pinnately compound, with 7 to 13 falcate leaflets, underside rusty scurfy and pubescent; buds yellow-brown, valvate; fruit a nut in a flattened, winged husk; bark gray and shaggy.

DESCRIPTION Leaves are alternate, pinnately compound, deciduous, and up to 40 cm (15.7 in) long, with 7 to 13 (sometimes 5 to 17) leaflets. Leaflets are lanceolate to falcate, 5–12 cm (2.0–4.7 in) long, and sessile; margin is finely serrate; underside is rusty scurfy and pubescent; autumn color is yellow. Twigs are red-brown, mottled, and glabrous or scurfy pubescent, with lenticels; leaf scar is heart shaped with numerous bundle scars. The terminal bud is up to 1 cm (0.4 in) long and conical; scales are valvate, yellow-brown, and scurfy. Flowers are imperfect and appear with the leaves in the spring; staminate flowers are in drooping catkins; pistillate flowers are in terminal spikes. Fruit is a nut, 2.5–4.0 cm (1.0–1.6 in) long, flattened, and ridged, with a bitter kernel; husk is thin with four thinly winged sutures that split to the base; fruit matures in the fall. Bark is gray-brown and smooth on

From left to right:
Water hickory leaf.
Water hickory fruit.

From left to right:

Water hickory bark.

Water hickory with very shaggy bark.

small trees; large trees have scaly or shaggy bark. The growth form is up to 30 m (100 ft) in height and 1 m (3 ft) in diameter.

HABITAT Swamps, bottomlands, wet clay flats, and shallow sloughs.

NOTES Water hickory is found growing with bitternut hickory, sugarberry, green ash, pumpkin ash, water locust, sweetgum, swamp tupelo, spruce pine, swamp laurel oak, overcup oak, cherrybark oak, Nutall oak, baldcypress, and American elm. The wood is white to redbrown, hard, and heavy, and suffers from longitudinal separations or cracks. The nuts are very bitter but are eaten by some wildlife.

Carya is derived from a Greek word meaning "walnut tree"; *aquatica* refers to the wet habitat.

Bitternut Hickory

Carya cordiformis (Wangenh.) K. Koch

COMMON NAMES bitternut hickory, pecan hickory, swamp hickory

QUICK GUIDE Leaves alternate, pinnately compound, with 7 to 11 leaflets; buds sulfur yellow, valvate; fruit husk with winged sutures above the middle; bark gray-brown with tight interlacing ridges forming a diamond pattern.

DESCRIPTION Leaves are alternate, pinnately compound, deciduous, and up to 25 cm (9.8 in) long, with 7 to 9 (sometimes 11) leaflets. Leaflets are obovate to lanceolate, 5–15 cm (2.0–5.9 in) long, and sessile; margin is serrate; underside is lightly pubescent; rachis lightly pubescent; autumn color is yellow. Twigs are gray-brown and lightly pubescent, with lenticels; leaf scar is heart shaped with numerous bundle scars. The terminal bud is about 1.5 cm (0.6 in) long, falcate, and flattened; scales are valvate, sulfur yellow, scurfy, and pubescent. Flowers are imperfect and appear with the leaves in the spring; staminate flowers are in drooping catkins; pistillate flowers are in terminal spikes. Fruit is a nut, nearly round and pointed, about 2.5 cm (1.0 in) long, and lightly ridged, with a bitter kernel; husk is thin, with four sutures that are thinly winged above the middle and split to the base; fruit matures in the fall. Bark is gray-brown and smooth on

Clockwise from left:

Bitternut hickory leaf.

Bitternut hickory fruit.

Bitternut hickory leaf form from underside; photo by Lisa J. Samuelson.

Bitternut hickory
bark.

small trees; bark on large trees is fissured with tight interlacing ridges forming a diamond pattern. The growth form is up to 30 m (100 ft) in height and 1 m (3 ft) in diameter.

HABITAT Moist to wet soils.

NOTES Bitternut hickory is found with shagbark hickory, sugarberry, numerous ashes, tulip-poplar, eastern cottonwood, spruce pine, white oak, northern red oak, bottomland oaks, basswood, and American elm. The wood is white to red-brown, hard, and heavy, and is used for pulpwood, handles, furniture, flooring, novelty items, ladders, skis, and fuel. The nuts are very bitter but are eaten by some wildlife.

Carya is derived from a Greek word meaning "walnut tree"; *cordiformis* means "heart shaped."

Pignut Hickory

Carya glabra (Mill.) Sweet

COMMON NAMES pignut hickory, sweet pignut, smooth hickory, white hickory

QUICK GUIDE Leaves alternate, pinnately compound, with five to seven leaflets, underside mostly glabrous, rachis glabrous; twigs slender and glabrous; fruit a nut, round or pear shaped, in a thin husk splitting halfway to the base; bark gray-brown with interlacing ridges forming a diamond pattern.

DESCRIPTION Leaves are alternate, pinnately compound, deciduous, and up to 30 cm (11.8 in) long, with usually five (sometimes seven) leaflets. Leaflets are obovate to lanceolate or falcate, 8–15 cm (3.1–5.9 in) long, and sessile; margin is serrate; underside is glabrous or sparsely pubescent; rachis slender and glabrous; autumn color is yellow. Twigs are slender, red-brown, and glabrous, with lenticels; leaf scar is heart shaped with numerous bundle scars. The terminal bud is up to 1 cm (0.4 in) long and ovoid; outer scales are red-brown, loose, and glabrous; inner scales are paler with silky hairs. Flowers are imperfect and appear with the leaves in the spring; staminate flowers are in drooping catkins; pistillate flowers are in terminal spikes. Fruit is a nut, nearly round to egg or pear shaped (like a pig's snout), 2.5–5.0 cm (1–2 in) long, and smooth, with a sweet kernel; husk is dark red-brown and thin and splits halfway to the base; fruit matures in the fall. Bark is smooth or lightly grooved on small trees; large trees are gray-brown and fissured with interlacing ridges forming a diamond pattern; bark

From left to right:

Pignut hickory leaf with five leaflets.

Pignut hickory leaf with seven leaflets, photo by Lisa J. Samuelson.

Pignut hickory fruit.

Pignut hickory bark.

becomes scaly on large trees. The growth form is up to 36 m (120 ft) in height and 1.2 m (4 ft) in diameter.

HABITAT Upland forests.

NOTES Forest associates of pignut hickory are numerous and include many upland oak, pine, and hickory species. The wood is white to red-brown, hard, and heavy, and is used for pulpwood, handles, furniture, flooring, novelty items, ladders, skis, and fuel. The wood is also used as smoker wood in barbeques. The nuts are eaten by white-tailed deer, wild turkey, squirrels, wood rats, chipmunk, bear, foxes, raccoon, crows, and woodpeckers.

Carya is derived from a Greek word meaning "walnut tree"; *glabra* means "hairless" and refers to the mostly hairless twig and leaf.

Pecan

Carya illinoinensis (Wangenh.) K. Koch

COMMON NAMES pecan, pecan hickory, sweet pecan

QUICK GUIDE Leaves alternate and pinnately compound, with 9 to 17 falcate leaflets; buds gray-brown, valvate, pubescent; fruit an ellipsoidal, sweet nut enclosed in a thin husk with thinly winged sutures; bark silver-gray, ridged, or scaly.

DESCRIPTION Leaves are alternate, pinnately compound, deciduous, and up to 50 cm (19.7 in) long, with 9 to 17 leaflets and a glabrous or pubescent rachis. Leaflets are lanceolate or falcate, 8–15 cm (3.1–5.9 in) long, and sessile; margin is serrate or doubly serrate; underside is pale pubescent; autumn color is yellow. Twigs are moderately stout, red-brown, and pubescent, with lenticels; leaf scar is heart shaped with numerous bundle scars. The terminal bud is ovoid and about 1 cm (0.4 in) long; scales are valvate, gray-brown to yellowish, and pubescent. Flowers are imperfect and appear with the leaves in the spring; staminate flowers are in drooping catkins; pistillate flowers are in terminal spikes. Fruit is a nut, ellipsoidal, 2.5–5.0 cm (1–2 in) long, and smooth or lightly ridged, with a sweet kernel; husk is thin with thinly winged sutures; fruit matures in the fall. Bark is gray-brown to silvery gray, shallowly ridged, and scaly, often with woodpecker holes. The growth form is up to 45 m (150 ft) in height, and old, open-grown trees may have a short trunk and wide spreading crown.

From left to right:

Pecan leaf.

Pecan fruit.

From left to right:

Pecan bark.

Pecan open-growth form; photo by Lisa J. Samuelson.

HABITAT Originally grew in well-drained bottomlands but now found on a variety of sites.

NOTES Pecan is the largest of the native hickories. Pecan has been planted extensively throughout the Southeast for nut production and subsequently its range has extended eastward and it is now found growing on a variety of sites. The wood is white to red-brown, hard, and heavy, and is used for pulpwood, handles, furniture, paneling, flooring, novelty items, and fuel. Pecan is a commercially important nut producer and popular ornamental despite its brittle limbs. Because of the sweet meat and thin shell, the nuts are favored by all squirrels, foxes, raccoon, and some bird species (I have seen crows drop the nuts on the road for cars to crush).

Carya is derived from a Greek word meaning "walnut tree"; *illinoinensis* refers to the geographic range.

Shellbark Hickory

Carya laciniosa (Michx. f.) G. Don

COMMON NAMES shellbark hickory, big shagbark hickory, bigleaf shagbark hickory, kingnut hickory

QUICK GUIDE Similar to shagbark hickory (*Carya ovata*) but leaves with seven larger leaflets; twigs orange-brown to yellow-brown; fruit a larger nut.

DESCRIPTION Leaves are alternate, pinnately compound, deciduous, and up to 60 cm (23.6 in) long, with seven (but sometimes five or nine) leaflets and a glabrous to densely pubescent rachis. Leaflets are obovate to oblanceolate, 5–22 cm (2.0–8.7 in) long, and sessile; margin is serrate and possibly ciliate; underside has a dense pale or yellow-maroon pubescence; autumn color is yellow. Twigs are stout, orange-brown to yellowish, and glabrous or pubescent, with lenticels; leaf scar is heart shaped with numerous bundle scars. The terminal bud is up to 3 cm (1.2 cm) long and ovoid; outer scales are pubescent and loose. Flowers are imperfect and appear with the leaves in the spring; staminate flowers are in drooping catkins; pistillate flowers are in terminal spikes. Fruit is a nut, egg shaped to nearly round, up to 6 cm (2.4 in) long (the largest of the hickories), and somewhat flattened, with four to six ribs and a sweet kernel; husk is depressed at the apex, brown, and very thick (1.2 cm [0.5 in]), with four and sometimes six deep sutures

Clockwise from left:

Shellbark hickory leaf.

Shellbark hickory nut and thick husk.

Shellbark hickory fruit.

Shellbark hickory bark.

splitting to the base and a pale interior; fruit matures in the fall. Bark on smaller trees has scaly ridges and on large trees has large, loose, shaggy plates. The growth form is up to 30 m (100 ft) in height and 1 m (3 ft) in diameter.

HABITAT Moist to periodically inundated soils, loamy bottoms, and occasionally on drier sites.

NOTES Shellbark hickory is an uncommon tree found in pure stands or with shagbark hickory, white ash, sweetgum, eastern cottonwood, swamp chestnut oak, Shumard oak, and American elm. The wood is white to red-brown, hard, heavy, and flexible, and is used for pulpwood, handles, furniture, sporting goods, novelty items, and fuel. The nuts are eaten by humans, squirrels, wild turkey, foxes, waterfowl, and small mammals.

Carya is derived from a Greek word meaning "walnut tree"; *laciniosa* is Latin for "shred or torn," referring to the shaggy bark.

Nutmeg Hickory

Carya myristiciformis (Mich. f) Elliott

COMMON NAMES nutmeg hickory, swamp hickory

QUICK GUIDE Leaves alternate, pinnately compound, with nine leaflets; underside with yellow or silvery scales; twigs and valvate buds scurfy; fruit with winged sutures; bark scaly on large trees.

DESCRIPTION Leaves are alternate, pinnately compound, deciduous, and up to 35 cm (13.8 in) long, usually with 9 (but sometimes 5 to 11) leaflets. Leaflets are obovate, 5–15 cm (2.0–5.9 in) long, and sessile; margin is serrate; petiole and rachis are scurfy; underside has yellow or silver scales that can appear shiny; autumn color is yellow. Twigs are brown-gray and scurfy; leaf scar is heart shaped with numerous bundle scars. The terminal bud is ovoid and up to 1 cm (0.4 in) long; scales are valvate, yellowish, and scurfy. Flowers are imperfect and appear with the leaves in the spring; staminate flowers are in drooping catkins; pistillate flowers are in terminal spikes. Fruit is a nut, ellipsoidal, and up to 4 cm (1.6 in) long, with a sweet kernel; husk is thin with yellow scales and four winged sutures that split almost to the base; fruit matures in the fall. Bark is red-brown to brown-gray and smooth on young trees and becomes scaly with loose plates on large trees. The growth form is up to 30 m (100 ft) in height.

From left to right:

Nutmeg hickory leaf and fruit.

Nutmeg hickory staminate flowers.

Nutmeg hickory fruit.

Bark of young
nutmeg hickory.

HABITAT Rich, moist soils of riverbanks, swamp margins, bottoms, and limestone prairies.

NOTES Nutmeg hickory is considered the rarest of hickories and is found growing with white ash, blackgum, swamp chestnut oak, cherrybark oak, Shumard oak, and other bottomland hickories. The sweet nuts are enjoyed by wildlife and used in cooking.

Carya is derived from a Greek word meaning "walnut tree;" *myristiciformis* refers to the similarity of the nut to nutmeg (*Myristica fragrans*).

Red Hickory

Carya ovalis (Wangenh.) Sarg.

COMMON NAMES red hickory, false hickory

QUICK GUIDE Similar to pignut hickory (*Carya glabra*), but leaves more commonly have seven rather than five leaflets and a red petiole; fruit with a rough husk splitting all the way to the base; bark shaggy.

DESCRIPTION Leaves are alternate, pinnately compound, deciduous, and up to 30 cm (11.8 in) long, with five to seven (sometimes nine) leaflets, a red petiole, and a slender and glabrous rachis. Leaflets are obovate to lanceolate or sometimes falcate, 8–15 cm (3.1–5.9 in) long, and sessile; margin is serrate; underside is glabrous or sparsely pubescent; autumn color is yellow. Twigs are slender, red-brown, and glabrous or lightly pubescent, with lenticels; leaf scar is heart shaped with numerous bundle scars. The terminal bud is ovoid and about 8 mm (0.3 in) long; inner scales are paler with silky hairs; outer scales are red-brown, loose, and glabrous or pubescent. Flowers are imperfect and appear with the leaves in the spring; staminate flowers are in drooping catkins; pistillate flowers are in terminal spikes. Fruit is a nut, nearly round, 2.5–4.0 cm (1.0–1.6 in) long, and unribbed, with a sweet kernel; husk is rough, dark red-brown, and thin, and splits to the base; fruit matures in the fall. Bark is smooth or lightly scaly on

From left to right:

Red hickory leaf and fruit.

Red hickory fruit.

Red hickory bark.

small trees; bark on large trees is gray-brown and fissured with scaly, loose, interlacing ridges. The growth form is up to 36 m (120 ft) in height and 1.2 m (4 ft) in diameter.

HABITAT Upland forests.

NOTES Red hickory is considered by some as a variety of pignut hickory, and red hickory often intergrades with pignut hickory. The wood is used for pulpwood, handles, furniture, flooring, novelty items, ladders, skis, and fuel. The wood is also used as smoker wood in barbeques. The nuts are eaten by white-tailed deer, wild turkey, squirrels, wood rats, chipmunk, bear, foxes, raccoon, crows, and woodpeckers.

Carya is derived from a Greek word meaning "walnut tree"; *ovalis* means "oval" or "egg shaped."

Shagbark Hickory

Carya ovata (Mill.) K. Koch

COMMON NAMES shagbark hickory, scalybark hickory, little shellbark hickory

QUICK GUIDE Leaves alternate, pinnately compound, usually with five leaflets, margin serrate and ciliate; fruit a large, round nut enclosed by a thick husk with four deep sutures splitting to the base; bark gray and shaggy with loose curved plates.

DESCRIPTION Leaves are alternate, pinnately compound, deciduous, and up to 40 cm (15.7 in) long, with five but sometimes seven leaflets and a stout and pubescent or glabrous rachis. Leaflets are variable in shape (from widely obovate with an abruptly acuminate apex to elliptical or lanceolate), 8–18 cm (3.1–7.1 in) long, and sessile; margin is serrate with fine tufts of hair; underside is glabrous or pubescent; autumn color is yellow. Twigs are stout, red-brown, and glabrous or pubescent, with lenticels; leaf scar is heart shaped to semicircular with numerous bundle scars. The terminal bud is up to 2 cm (0.8 in) long and ovoid; outer scales are dark, pubescent, and very loose; inner scales are paler with silky hairs. Flowers are imperfect and appear with the leaves in the spring; staminate flowers are in drooping catkins; pistillate flowers are in terminal spikes. Fruit is a nut, egg shaped or nearly round, up to 4 cm (1.6 in) long, and white to light brown, with four ridges and a sweet kernel; husk is depressed at the apex, brown to black, shiny, and very thick (12 mm), with four deep sutures that split to the base and a pale interior; fruit matures in the fall. Bark is smooth or lightly

Clockwise from left:

Shagbark hickory leaf.

Shagbark hickory ciliate leaflet margin.

Shagbark hickory nut and thick husk.

Shagbark hickory fruit.

Shagbark hickory
bark.

grooved on small trees; large trees are gray and shaggy with curved, loose plates. The growth form is up to 39 m (130 ft) in height and 1.2 m (4 ft) in diameter.

HABITAT Moist upland sites, swamp edges, and in well-drained bottomlands.

NOTES Common associates of shagbark hickory include tulip-poplar, sugar maple, eastern white pine, white oak, northern red oak, and black oak on upland sites, and in bottoms bitternut hickory, boxelder, white ash, sweetgum, bottomland oaks, and sugarberry. The wood is white to red-brown, hard, and heavy, and is used for pulpwood, handles, furniture, paneling, flooring, novelty items, and fuel. The nuts are eaten by waterfowl, black bear, foxes, squirrels, and other small mammals.

Carya is derived from a Greek word meaning "walnut tree"; *ovata* means "egg shaped."

SIMILAR SPECIES **Southern shagbark hickory** (*Carya carolinae-septentrionalis* [Ashe] Engl. & Graebn.) is found in bottoms and moist woods and on limestone soils. It is distinguished by smaller buds with shiny reddish or blackish outer scales, smaller fruit (nut up to 3.0 cm [1.2 in] long, husk 3–9 mm [0.1–0.4 in] thick), and slender black twigs.

Sand Hickory

Carya pallida (Ashe) Engl. & Graebn.

COMMON NAMES sand hickory, pale-leaf hickory, pale hickory

QUICK GUIDE Leaves alternate, pinnately compound, with five to nine leaflets, underside pale with pubescence and silvery scales; fruit a round nut enclosed in husk with yellow scales, splitting to the base; bark brown-gray with interlacing ridges.

DESCRIPTION Leaves are alternate, pinnately compound, deciduous, and up to 40 cm (15.7 in) long, with five to nine (but usually seven) leaflets and a slender and pubescent rachis. Leaflets are obovate to lanceolate or falcate, 8–14 cm (3.1–5.5 in) long, and sessile; margin is serrate; underside is pale and has silvery scales and pubescence; autumn color is yellow. Young leaves have silvery scales on the surface. Twigs are slender and red-brown, with yellow-brown scales and lenticels; leaf scar is heart shaped with numerous bundle scars. The terminal bud is about 8 mm (0.3 in) long and ovoid; scales are yellow, overlapping, scurfy, and fringed with pubescence. Flowers are imperfect and appear with the leaves in the spring; staminate flowers are in drooping catkins; pistillate flowers are in terminal spikes. Fruit is nut, nearly round to egg shaped, up to 4 cm (1.6 in) long, and ridged, with a sweet kernel; husk has yellow scales and splits to the base; fruit

Clockwise from left:

Sand hickory leaf with seven leaflets.

Sand hickory fruit.

Sand hickory leaf, with five leaflets and pale underside, and the fruit; photo by Lisa J. Samuelson.

Sand hickory bark.

matures in the fall. Bark is smooth or lightly grooved on small trees; bark on large trees is gray-brown and fissured with lightly scaly interlacing ridges. The growth form is up to 30 m (100 ft) tall.

HABITAT Upland forests on ridges, sandy soils, and dry slopes.

NOTES In the sandy soils of the Southern Coastal Plain, sand hickory is found with longleaf pine, bluejack oak, turkey oak, sand post oak, and blackjack oak. In the Piedmont, forest associates include pignut hickory, mockernut hickory, scarlet oak, and post oak. The wood is white to red-brown, hard, and heavy, and is used primarily for fuel. The nuts are eaten by squirrels and other small mammals.

Carya is derived from a Greek word meaning "walnut tree"; *pallida* means "pale" and refers to the leaf underside.

Mockernut Hickory

Carya tomentosa (Lam. ex Poir.) Nutt.

COMMON NAMES mockernut hickory, big bud hickory, white hickory

QUICK GUIDE Leaves alternate, pinnately compound, with seven to nine leaflets, underside and rachis pubescent or tomentose; twigs stout and hairy; fruit a large nut with a thick husk that splits to the base; bark gray-brown with interlacing ridges.

DESCRIPTION Leaves are alternate, pinnately compound, deciduous, and up to 43 cm (16.9 in) long, with seven to nine (sometimes five) leaflets and a stout and pubescent rachis. Leaflets are obovate to oblanceolate or lanceolate, 5–18 cm (2.0–7.1 in) long, and sessile; margin is serrate; underside is pubescent or rusty tomentose; autumn color is yellow. Twigs are stout, red-brown, and tomentose or pubescent, with lenticels; leaf scar is heart shaped with numerous bundle scars. The terminal bud is up to 2 cm (0.8 in) long and ovoid; outer scales are tomentose and loose; inner scales are paler with silky hairs. Flowers are imperfect and appear with the leaves in the spring; staminate flowers are in drooping catkins; pistillate flowers are in terminal spikes. Fruit is an elliptical to nearly round nut, up to 5 cm (2 in) long, and four

Clockwise from left:

Mockernut hickory leaf.

Mockernut hickory fruit.

Mockernut hickory staminate flowers.

From left to right:

Bark of young mockernut hickory.

Mockernut hickory bark.

ridged, with a sweet kernel; husk is thick (5 mm [0.2 in]), dark red-brown, and shiny, with four deep sutures that split to the base; fruit matures in the fall. Bark is blue-gray and smooth or lightly grooved on small trees; bark of large trees is gray-brown and fissured with interlacing ridges forming a diamond pattern. The growth form is up to 30 m (100 ft) in height and 1 m (3 ft) in diameter.

HABITAT A variety of sites from bottomlands to fertile uplands and dry rocky slopes.

NOTES Forest associates of mockernut hickory are numerous and site dependent. On upland sites mockernut hickory is found with pignut hickory, eastern redcedar, blackgum, sourwood, shortleaf pine, loblolly pine, upland oaks, and winged elm. The wood is white to red-brown, hard, and heavy, and is used for pulpwood, handles, furniture, paneling, flooring, novelty items, and fuel. The nuts are a favorite of squirrels and other small mammals, and are also eaten by white-tailed deer, foxes, and beaver.

Carya is derived from a Greek word meaning "walnut tree"; *tomentosa* refers to hair on the leaves and twigs. Some texts refer to this species as *Carya alba*.

Butternut

Juglans cinerea L.

COMMON NAMES butternut, white walnut, oil nut

QUICK GUIDE Leaves alternate, pinnately compound, with 11 to 17 ovate leaflets, usually with a terminal leaflet, petiole and rachis sticky hairy; leaf scar with a fuzzy mustache; fruit an ellipsoidal, prominently ridged, corrugated nut encased in a sticky pubescent husk; bark gray and grooved with flattened ridges.

DESCRIPTION Leaves are alternate, pinnately compound, deciduous, and up to 75 cm (29.5 in) long, with 11 to 17 leaflets, usually a terminal leaflet, and sticky hair on the petiole and rachis. Leaflets are ovate to oblong-lanceolate, 5–12 cm (2.0–4.7 in) long, and nearly sessile; margin is serrate; underside is pale and pubescent; autumn color is yellow. Twigs are stout, brown, and glabrous or pubescent, with lenticels; pith is dark chocolate brown and chambered; leaf scar is three-lobed with a hairy fringe or "mustache" above the scar, and bundle scars are in groups of three (forming a monkey face). The terminal bud is elongated, conical, and about 1.5 cm (0.6 in) long; scales are overlapping, white or gray-brown, scurfy, and pubescent; lateral buds are superposed. Flowers are imperfect and appear after the leaves in the spring;

Clockwise from left:
Butternut leaf.
Butternut nut.
Butternut fruit.

staminate flowers are in drooping catkins; pistillate flowers are in terminal spikes. Fruit is a nut, ellipsoid to oblong or ovoid, about 6 cm
(2.4 in) long, corrugated, and sharply ridged, with an oily and sweet
kernel; husk is yellow-green to green-brown, pubescent, sticky, thick,
and indehiscent; fruit matures in the fall. Bark is ash gray with interlacing ridges on small trees; large trees have flat white ridges; inner
bark is chocolate brown. The growth form is up to 30 m (100 ft) in
height and 1 m (3 ft) in diameter.

HABITAT Moist fertile soils such as stream banks, upland slopes, and
in coves.

NOTES The distribution of butternut has been greatly reduced due to
butternut canker, a fungus that causes stem girdling cankers. The
wood is white to chestnut brown, coarse-grained, light, and soft, and
is used for plywood, boxes, trim work, cabinets, and furniture. A dye
was made from the husks and bark. The nuts are eaten by people and
immature nuts were pickled to make an oilnut relish.

Juglans is Latin for "walnut" (glans meaning acorn or nut); *cinerea*
means "ash," referring to the bark.

Black Walnut

Juglans nigra L.

COMMON NAMES black walnut, American walnut

QUICK GUIDE Leaves alternate, pinnately compound, with 8 to 24 leaflets; leaflets ovate, often lacking the terminal leaflet; fruit a round, ridged, corrugated nut encased in a yellow-green husk; bark dark brown and furrowed with interlacing ridges.

DESCRIPTION Leaves are alternate, pinnately compound, deciduous, and up to 60 cm (23.6 in) long, with 8 to 24 leaflets and lightly pubescent rachis. Leaflets are ovate to ovate-lanceolate, 5–13 cm (2.0–5.1 cm) long, and nearly sessile; margin is serrate; underside is pubescent; autumn color is yellow. Mature trees often have an aborted or malformed terminal leaflet whereas young trees often possess a terminal leaflet. Twigs are stout, brown, and pubescent, with lenticels; pith is light brown and chambered; leaf scar is three-lobed with bundle scars in groups of three that form a monkey face. The terminal bud is conical to round and about 1 cm (0.4 in) long; scales are overlapping, brown-gray, and pubescent; lateral buds are superposed. Flowers are imperfect and appear after the leaves in the spring; staminate flowers

Clockwise from left:
Black walnut leaf.
Black walnut nut.
Black walnut fruit.

Black walnut bark.

are in drooping catkins; pistillate flowers are in terminal spikes. Fruit is a nut, round, up to 6 cm (2.4 in) wide, weakly ridged, and corrugated, with a sweet kernel; husk is yellow-green, indehiscent, and thick; fruit matures in the fall. Bark is brown to dark brown and furrowed, with interlacing ridges; inner bark is chocolate brown. The growth form is up to 30 m (100 ft) in height.

HABITAT Moist, fertile soils.

NOTES Black walnut is found growing with bitternut hickory, shagbark hickory, American beech, white ash, tulip-poplar, black cherry, white oak, northern red oak, basswood, and American elm. The wood has white sapwood and red-brown heartwood and is dense, fine-textured, strong, and hard. The wood is commercially valuable and is used for furniture, cabinetry, veneer, and rifles. The nuts are eaten by foxes, squirrels, and some birds. Beaver will occasionally use the bark.

Juglans is Latin for "walnut tree," *nigra* means "black," referring to the bark or wood.

Laurel Family (*Lauraceae*)

Redbay

Persea borbonia (L.) Spreng.

COMMON NAMES redbay

QUICK GUIDE Leaves alternate, simple, elliptical, often disfigured from insect galls, evergreen, with a spicy aroma when crushed, underside pale; fruit a round blue drupe; bark gray-brown to red-brown with flattened ridges.

DESCRIPTION Leaves are alternate, simple, evergreen, elliptical to oblong or oblanceolate, often disfigured from insect galls, 7–15 cm (2.8–5.9 in) long, and leathery; apex and base are acute to rounded; margin is entire; petiole is stout; underside is pale with minute pubescent; leaves have a spicy aroma when crushed. Twigs are slender, slightly angled, mottled green-brown, and pubescent; leaf scar is crescent shaped with one linear bundle scar. The terminal bud is naked, rusty pubescent, and about 5 mm (0.2 in) long. Flowers are perfect, yellow-white, and about 5 mm (0.2 in) wide, and bloom in the spring with the new leaves. Fruit is a blue drupe, nearly round, and about 1 cm (0.4 in) wide, and matures in early autumn. Bark is gray-brown to red-brown, with flattened ridges; large trees are more furrowed; inner bark is red-brown. The growth form is up to 18 m (60 ft) tall and 60 cm (2 ft) in diameter.

Clockwise from left:

Redbay leaves; photo by Lisa J. Samuelson.

Redbay leaves showing insect galls.

Redbay fruit.

Bark of young redbay.

HABITAT Forested wetlands, pocosins, swamps, mesic slopes, and long-leaf pine woodlands, as well as on drier sites.

NOTES Forest associates of redbay include sweetbay magnolia, swamp tupelo, Atlantic white-cedar, sweetgum, baldcypress, loblolly-bay, slash pine, and pond pine. The wood is red, heavy, and hard, and is used for furniture, cabinets, boats, and trim work. The leaves were used as a substitute for bay leaves in cooking. Deer may browse the foliage and eat the fruit. The fruit is also eaten by songbirds, northern bobwhite, wild turkey, and small mammals.

Persea is an ancient Greek name for an Oriental tree; *borbonia* is an old name for Persea.

SIMILAR SPECIES Swamp redbay (*Persea palustris* [Raf.] Sarg.) is a smaller tree found on wetter sites such as swamps and bays. It is identified by leaves with rusty hairs on the midrib and veins, and twigs with dense, long brown hairs.

Sassafras

Sassafras albidum (Nutt.) Nees

COMMON NAMES sassafras, cinnamon wood

QUICK GUIDE Leaves alternate, simple, elliptical, margin entire, or with two or three lobes; leaves, twigs, and bark aromatic when cut; twigs mottled yellow-green; fruit a dark blue, red-stalked drupe; bark brown and thickly ridged.

DESCRIPTION Leaves are alternate, simple, deciduous, elliptical to ovate or obovate or oval, 8–15 cm (3.1–5.9 in) long, and aromatic when crushed; margin is entire or two- or three-lobed; veins are prominent on the upper surface; underside is pale with pubescence along the veins; autumn color is red, orange, or yellow. Twigs are slender, yellow-green, mottled, often curved, aromatic when cut, and finely pubescent, with lenticels; leaf scar is half-round with one linear bundle scar. The terminal bud is ovoid, yellow-green, and up to 1 cm (0.4 in) long, with three to four overlapping and keeled scales. Flowers are dioecious and bloom in the spring usually before but sometimes with the leaves in yellow clusters on branch ends. Fruit is a drupe, dark blue, oblong to nearly round, up to 1 cm (0.4 in) long, and born on bright-red swollen stalks in early fall. Bark is gray to red-brown, thick,

Clockwise from upper left:

Sassafras leaves.

Sassafras tree in flower.

Sassafras fruit.

Sassafras bark.

and irregularly fissured, with broad ridges; inner bark is red-brown and fragrant when cut. The growth form is up to 15 m (50 ft) in height but can be larger and can form thickets.

HABITAT Well-drained soils.

NOTES Sassafras is usually a small tree found growing on a variety of sites but can become a larger tree on better sites. The sapwood is yellow, and the heartwood is orange-brown. The wood is moderately heavy and moderately hard, and is used for furniture, specialty items, millwork, window frames, cabinets, and fence posts. In the past, the wood was used for fishing rods, boat construction, and ox yokes. Oil of sassafras is extracted from the roots for perfumes and teas. The leaves are used in flavoring gumbo. Sassafras is a browse for white-tailed deer, black bear, beaver, rabbits, and woodchuck. The fruit is of moderate value to a variety of birds and mammals.

Sassafras comes from a Native America name; *albidum* means "whitish," referring to the leaf underside.

Magnolia Family (*Magnoliaceae*)

Tulip-Poplar

Liriodendron tulipifera L.

COMMON NAMES tulip-poplar, yellow-poplar, tuliptree, white-poplar, whitewood

QUICK GUIDE Leaves broadly lobed, alternate, simple, tulip shaped; bud shaped like a duck's bill; flower tuliplike; petals yellow-green with orange; fruit a persistent cone of samaras; bark ash gray and deeply furrowed.

DESCRIPTION Leaves are alternate, simple, deciduous, and 13–25 cm (5.1–9.8 in) long, mostly with four broad lobes; apex is broadly notched to truncate; base is truncate; margin is smooth; petiole is long; autumn color is bright yellow. Twigs are moderately stout, red-brown to purple-maroon, and glabrous, with stipule scars encircling the twig; leaf scar is round and raised, with numerous bundle scars. The terminal bud is yellow-green to dark purple, up to 2 cm (0.8 in) long, and flattened, with two valvate scales, like a duck's bill. Flowers are perfect, tuliplike, and up to 10 cm (3.9 in) wide, and bloom in late spring or early summer; petals are yellow-green with orange at the base. Fruit is a samara, about 3.5 cm (1.4 in) long, and in a conelike aggregation up to 8 cm (3.1 in) long; fruit matures in early autumn. Samara are shed throughout the winter and open cones persist on branches. Bark is gray or gray-green and smooth with light vertical grooves and black

From left to right:

Tulip-poplar leaves and flower.

Tulip-poplar fruit.

From left to right:

Bark of young tulip-poplar.

Bark of large tulip-poplar.

branch scars on small trees; large trees are ash gray, thick, and deeply furrowed. The growth form is a tall tree up to 46 m (150 ft) in height and 1.5 m (5 ft) in diameter with a straight, clear bole.

HABITAT Moist, fertile soils, such as stream banks, cool slopes, well-drained bottomlands, coves, and ravines.

NOTES Tulip-poplar is one of the tallest trees in North America and fast-growing on good sites. Forest associates include red maple, sugar maple, common persimmon, American beech, sweetgum, sweetbay magnolia, cucumbertree, southern magnolia, blackgum, eastern hemlock, shortleaf pine, loblolly pine, black cherry, white oak, northern red oak, black oak, and eastern hemlock. The wood is white to green-brown with green to black stripes and is light, soft, and straight-grained. The wood is used for pulpwood, veneer, furniture, paneling, trim, framing, cabinets, pallets, musical instruments, and boxes. Tulip-poplar can be planted as a shade tree, but it requires a large area, plenty of sun, and moist soil. The flowers are very popular with bees. Tulip-poplar is a preferred white-tailed deer browse when succulent in the spring and summer. The seeds are eaten by a wide variety of songbirds and small mammals. Beaver will cut young stems growing near water.

Liriodendron means "lily-tree"; *tulipifera* means "tuliplike."

Cucumbertree

Magnolia acuminata (L.) L.

COMMON NAMES cucumbertree, cucumber magnolia, yellow-flower magnolia, mountain magnolia

QUICK GUIDE Leaves alternate, simple, elliptical, underside with silky pubescence; bud with white silky pubescence; flower yellow-green; fruit cone-like, bulbous, often curved like a cucumber, with shiny red follicles; bark red-brown to gray-brown with scaly ridges.

DESCRIPTION Leaves are alternate, simple, deciduous, elliptical to ovate or obovate to oval, and 13–25 cm (5.1–9.8 in) long; apex is acute to acuminate; base is acute to rounded; margin is entire; underside has soft white pubescence; autumn color is yellow. Twigs are moderately stout, red-brown, and glabrous, and stipule scars encircle the twig; leaf scar is U-shaped with numerous bundle scars. The terminal bud is ovoid and about 1.5 cm (0.6 in) long, with a single scale that is covered with silky white hairs. Flowers are perfect and 6–8 cm (2.4–3.1 in) long, with yellow-green petals, and bloom in late spring or early summer. Fruit is a follicle, about 1.3 cm (0.5 in) long, red, and shiny; follicles are in a bulbous, often curved, conelike aggregation up to 8 cm (3.1 in) long; fruit matures in early fall. Green immature fruit looks like a cucumber. Bark is red-brown to gray-brown with flattened, scaly ridges. The growth form is up to 30 m (100 ft) in height and 1.2 m (4 ft) in diameter.

From left to right:

Cucumbertree leaves.

Cucumbertree fruit with stinkbug on it.

Cucumbertree flower.

Cucumbertree
bark.

HABITAT Moist, deep soils of bottomlands, coves, and cool slopes.

NOTES Cucumbertree can be found growing with red maple, sugar maple, yellow buckeye, black birch, ashes, Carolina silverbell, butternut, tulip-poplar, other magnolias, eastern white pine, black cherry, white oak, northern red oak, basswood, eastern hemlock, and American elm. The wood is sold as tulip-poplar and is white to pale green-brown, heavy, hard, and straight-grained, and is used for pulpwood, veneer, furniture, paneling, trim, framing, pallets, and boxes. Overall use of the magnolias by wildlife is relatively low, but the seeds are eaten by some birds, such as cardinals, and small mammals.

Magnolia refers to the botanist P. Magnol; *acuminata* refers to the leaf apex.

Southern Magnolia

Magnolia grandiflora L.

COMMON NAMES southern magnolia, great laurel magnolia, evergreen magnolia, bull bay

QUICK GUIDE Leaves alternate, simple, evergreen, elliptical, leathery, underside with rusty pubescence; flower large, white, very fragrant; fruit conelike with shiny red follicles; bark brown and smooth or scaly.

DESCRIPTION Leaves are alternate, simple, evergreen, elliptical to oblong or oval, 13–25 cm (5.1–9.8 in) long, and leathery; apex is acute to rounded; base is acute to obtuse; margin is entire; upper surface is shiny dark green; underside has rusty pubescence. Twigs are stout and red-brown, with rusty or white hair and stipule scars encircling the twig; leaf scar is nearly round with numerous bundle scars. The terminal vegetative bud is up to 4 cm (1.6 in) long and conical, with one yellow-green scale with rusty hair; flower bud is covered with pale wooly hair. Flowers are perfect, up to 20 cm (7.9 in) wide, very fragrant, and very showy, with large white petals, and bloom in late spring or early summer. Fruit is a follicle, about 1.3 cm (0.5 in) long, and shiny bright red; follicles are in a conelike aggregation up to 10 cm (3.9 in) long; fruit matures in the fall. Bark is brown to green-brown, smooth or with loose flakes; large trees are scaly. The growth form is up to 24 m (80 ft) in height and 1 m (3 ft) in diameter.

From left to right:

Southern magnolia leaves.

Southern magnolia flower.

Southern magnolia fruit.

Southern magnolia smooth bark.

HABITAT Moist soils, such as bottomlands, stream and swamp edges, and low uplands.

NOTES Southern magnolia is a shade-tolerant species found with American beech, sweetgum, tulip-poplar, sweetbay magnolia, swamp tupelo, loblolly pine, cherrybark oak, and live oak. The wood is sold as tulip-poplar and is white to a dark green-brown, heavy, hard, and straight-grained, and is used for pulpwood, veneer, furniture, paneling, trim, framing, pallets, and boxes. Overall use of the magnolias by wildlife is relatively low, but the seeds are eaten by some birds (in particular cardinals) and small mammals (including gray squirrel). The dense evergreen foliage offers roosting, loafing, and nesting cover for a variety of birds. Honey bees visit the flowers. A popular ornamental because of the evergreen foliage, wonderfully fragrant flowers, and bright red fruit, but the fruit and leaves can be messy and the roots may invade septic lines.

Magnolia refers to the botanist P. Magnol; *grandiflora* refers to the large, showy flowers.

Bigleaf Magnolia

Magnolia macrophylla Michx.

COMMON NAMES bigleaf magnolia, large-leaved cucumbertree

QUICK GUIDE Leaves alternate, simple, up to 80 cm (31.5 in) long, with an earlike base, underside with white pubescence; flowers large and white; fruit conelike with shiny red follicles; bark gray and smooth with corky lenticels.

DESCRIPTION Leaves are alternate, simple, deciduous, mostly obovate, and 50–80 cm (19.7–31.5 in) long; apex is acute; base is auriculate; margin is entire; underside has white pubescence; autumn color is yellow. Twigs are stout, mottled green-black, and pubescent, and stipule scars encircle the twig; leaf scar is nearly round with numerous bundle scars. The terminal bud is elongated, flattened, and up to 6 cm (2.4 in) long, with one bud scale that is covered with pale pubescence. Flowers are perfect and up to 40 cm (15.7 in) wide; petals are creamy white with a purple blotch at the base; flowers bloom in the spring after the leaves. Fruit is a follicle, about 1.3 cm (0.5 in) long and shiny red; follicles are in a conelike, cylindrical aggregation up to 8 cm (3.1 in) long; fruit usually matures in early fall. Bark is light gray-brown and smooth with corky lenticels or warts; large trees are scaly at the base. The growth form is up to 15 m (50 ft) in height.

From left to right:

Bigleaf magnolia leaves.

Bigleaf magnolia flower.

Bigleaf magnolia fruit.

Bigleaf magnolia
bark.

HABITAT Moist, well-drained soils, such as coves and stream banks.

NOTES Bigleaf magnolia is found with Florida maple, pawpaw, river birch, tulip-poplar, sycamore, and basswood. The wood is similar to southern magnolia but is not commercially important due to limited availability. Overall use of the magnolias by wildlife is relatively low, but the seeds are eaten by some birds and small mammals. This tree is often planted as an ornamental for the large leaves and flowers.

Magnolia refers to the botanist P. Magnol; *macrophylla* means "large leaf."

Umbrella Magnolia

Magnolia tripetala (L.) L.

COMMON NAMES umbrella magnolia, umbrella tree

QUICK GUIDE Leaves alternate, simple, obovate, clustered at branch ends (like an umbrella), base acute, underside with soft white pubescence; bud glabrous with a single green-purple scale; flower creamy white with an unpleasant fragrance; fruit conelike with shiny red follicles; bark gray-brown and mostly smooth.

DESCRIPTION Leaves are alternate, simple, deciduous, obovate, and 25–60 cm (9.8–23.6 in) long; apex is acute; base is acute to cuneate; margin is entire; underside has soft white pubescence; autumn color is yellow. Leaves are clustered at branch ends like an umbrella. Twigs are moderately stout, green-brown to purplish, and mostly glabrous, with stipule scars that encircle the twig; leaf scar is nearly round with numerous bundle scars. The terminal bud is elongated, conical, curved, and up to 5 cm (2 in) long, with one green-purple, glabrous scale. Flowers are perfect and 20 cm (7.9 in) wide, with creamy white inner petals and an unpleasant fragrance, and bloom in the spring after the leaves. Fruit is a follicle, about 1.3 cm (0.5 in) long, and shiny red; follicles are in a conelike aggregation up to 12 cm (5.7 in) long; fruit matures in early fall. Bark is mottled gray-brown, and smooth, with lenticels and warts. The growth form is up to 12 m (40 ft) in height, and branches tend to curve upward, emphasizing the umbrella appearance.

HABITAT Coves, ravines, cool slopes, and stream edges.

From left to right:

Umbrella magnolia leaves; courtesy of Alan Cressler.

Underside of umbrella magnolia leaves.

From left to right:

Umbrella magnolia
fruit.

Umbrella magnolia
bark.

NOTES Forest associates of umbrella magnolia include cucumber-tree, bigleaf magnolia, yellow buckeye, Carolina buckthorn, Carolina silverbell, black walnut, white ash, and mesic site oaks. The wood is not commercially important. Overall use of the magnolias by wildlife is relatively low, but the seeds are eaten by some birds and small mammals.

Magnolia refers to the botanist P. Magnol, *tripetala* means "three-petalled," referring to the three outer sepals on the flower.

Sweetbay Magnolia

Magnolia virginiana L.

COMMON NAMES sweetbay magnolia, white bay, swamp bay, swamp magnolia

QUICK GUIDE Leaves alternate, simple, elliptical, underside prominently silvery blue; flowers with creamy white petals and fragrant; fruit conelike with shiny red follicles; bark mottled gray-brown and smooth.

DESCRIPTION Leaves are alternate, simple, semi-evergreen or evergreen, elliptical to oblong, 8–17 cm (3.1–6.7 in) long, leathery, and aromatic when crushed; apex is acute to obtuse; base is acute to cuneate; margin is entire; upper surface is green and shiny; underside is silvery blue-white. Twigs are slender and mottled green, with silky pubescence and stipule scars that encircle the twig; leaf scar is nearly round with numerous bundle scars. The terminal bud is up to 2 cm (0.8 in) long, conical, and flattened, with one green scale that is covered with silky, white-gray pubescence. Flowers are perfect, up to 8 cm (3.1 in) wide, and fragrant, with creamy white petals, and they bloom in late spring or early summer. Fruit is a follicle, about 1.3 cm (0.5 in) long, shiny, dark red; follicles are in a conelike aggregation up to 5 cm (2 in) long; fruit matures in early fall. Bark is mottled gray-brown and smooth, with lenticels; large trees are scaly at the base. The growth form is up to 30 m (100 ft) in height and 1 m (3 ft) in diameter.

Clockwise from left:

Silvery underside of sweetbay magnolia leaves.

Sweetbay magnolia flower.

Sweetbay magnolia fruit.

Sweetbay magnolia bark.

HABITAT Moist to wet but not deeply flooded soils of stream edges, swamps, and bottomlands.

NOTES Sweetbay magnolia is found with many species, including red maple, Atlantic white-cedar, American beech, southern magnolia, loblolly-bay, sweetgum, swamp tupelo, slash pine, pond pine, spruce pine, and baldcypress. The wood is white to pale green-brown, soft, and straight-grained, and is used for pulpwood, veneer, pallets and boxes. Overall use of the magnolias by wildlife is relatively low, but the seeds are eaten by small mammals and some birds. The bark is eaten by beaver.

Magnolia refers to the botanist P. Magnol; *virginiana* refers to the geographic range.

Hibiscus or Mallow Family (*Malvaceae*)

Basswood

Tilia americana L. var. *heterophylla* (Vent.) Loudon

Tilia americana L. var. *caroliniana* (Mill.) Castigl.

COMMON NAMES mountain basswood and white basswood (*heterophylla* variety); southern basswood and Carolina basswood (*carolininana* variety)

QUICK GUIDE Leaves alternate, simple, asymmetrically heart shaped, margin sharply serrate; fruit a nutlet attached to a leafy bract; bark gray-brown with flattened ridges. Southern basswood is found in the lower Southern Coastal Plain and has leaves with a gray or brown underside and tomentose hair, and lateral buds 3–5 mm (0.1–0.2 in) long. Mountain basswood is found in southern Appalachian forests and the Piedmont and has leaves with a pale or white underside and stellate tomentose hairs, and larger (5–8 mm [0.2–0.3 in] long) lateral buds.

DESCRIPTION Leaves are alternate, simple, deciduous, asymmetrically ovate to heart shaped, and 10–20 cm (3.9–7.9 in) long; apex is acuminate; base is cordate and inequilateral; margin is sharply serrate; underside is glabrous or pubescent; autumn color is yellow. Twigs zigzag and are moderately stout and green to red-brown; leaf scar is semicircular with numerous bundle scars. A true terminal bud is lacking;

From left to right:

Basswood leaves.

Basswood flowers.

Basswood fruit.

Bark of a large basswood tree.

lateral bud is divergent, ovoid, and plump; scales are two to three, green or maroon, and overlapping. Flowers are perfect, yellowish, fragrant, long-stalked, and attached to a leafy bract; flowers bloom in late spring or early summer. Fruit is a nutlet, 5–8 mm (0.2–0.3 in) wide, pubescent, long-stalked, and attached to a leafy bract; fruit matures in late summer or early fall. Bark is gray-brown and smooth or shallowly grooved on small trees; large trees are more deeply furrowed with flattened or somewhat scaly fibrous ridges. The growth form is up to 36 m (120 ft) tall and 1.2 m (4 ft) in diameter.

HABITAT The northern variety is found on moist, fertile soils, and the Southern Coastal Plain variety is found in mesic forests on sandy or limestone soils.

NOTES The taxonomy of *Tilia* is confused and some authors consider all varieties as one species. Basswood is found in association with many tree species. The wood is white to yellow-brown, light, soft, and fine textured, and is used for plywood, furniture, crates, boxes, turnery, guitars, piano keys, and woodenware. The fibrous inner bark was used for making ropes, fishnets, and mats. Basswood can be planted as an ornamental and cultivars are available. The flowers are very popular with honey bees, and their nectar is the source of basswood honey. The foliage is lightly browsed by white-tailed deer and rabbit, and seeds are eaten by squirrel and other small mammals. The dried leaves were used as winter fodder for cattle.

Tilia is Latin for "linden tree"; *americana* refers to the New World.

Mahogany Family (*Meliaceae*)

Chinaberry

Melia azedarach L.

COMMON NAMES chinaberry, Chinese umbrella tree, Pride of India

QUICK GUIDE Leaves alternate, bi- or tripinnately compound; leaflets coarsely serrate and some with one or two lobes at the base; flowers in purple fragrant clusters in the spring; fruit a yellow drupe in long-stalked clusters persisting over winter; bark brown-gray with loose interlacing ridges.

DESCRIPTION Leaves are alternate, bi- or tripinnately compound, deciduous, and up to 50 cm (19.7 in) long, with many leaflets. Leaflets are ovate to obovate and 3–8 cm (1.2–3.1 in) long; apex is acuminate; margin is coarsely serrate and sometimes has one or two lobes at the base; autumn color is yellow. Twigs are stout, purple-brown to green-brown, with brown pubescence and orange lenticels; leaf scar is raised and three-lobed, with three bundle scars. A true terminal bud is lacking; lateral bud is round, fuzzy, and covered with dense buff colored hairs. Flowers are perfect and fragrant, and bloom in pink-purple clusters in the spring after the leaves; individual flowers resemble purple firecrackers. Fruit is a drupe, round, 1–2 cm (0.4–0.8 in) wide, yellow, and leathery; fruit matures in early fall but persists over the

From left to right:

Chinaberry leaf.

Chinaberry flowers.

Chinaberry fruit.

winter. Bark on small stems is purple-green to maroon, shiny, and smooth, with lenticels; on large trees the bark is brown-gray with thick, loose, interlacing ridges. The growth form is up to 15 m (50 ft) in height with an open low crown.

HABITAT Originally from Asia but naturalized in the South. Found in thickets, in open fields, near old homesteads, and along fencerows, forest edges, and roadsides.

NOTES Chinaberry was once a popular ornamental tree and prized by children for the low branches, which were suitable for tree houses. The tree has become an invasive species that is difficult to control. The fruit is poisonous to livestock but is eaten by some birds, in particular cedar waxwings and American robins. The leaves were once used in bedding to repel fleas and lice, and the seeds were used as beads. The fruits are a favorite for sling shots and popguns.

Melia means "like an ash tree"; *azedarach* means "noble tree."

Fig Family (*Moraceae*)

Osage-Orange

Maclura pomifera (Raf.) C. K. Schneid.

COMMON NAMES Osage-orange, horse-apple, bow-wood, mock-orange, hedge-apple, bodark, bois d'arc

QUICK GUIDE Leaves alternate, simple, ovate, shiny, apex acuminate, petiole with white sap; twigs often with a stout spine at the node; fruit a large, green, brainlike ball of drupes; bark gray-brown and thick with interlacing ridges, inner bark orange.

DESCRIPTION Leaves are alternate, simple, deciduous, ovate, 8–15 cm (3.1–5.9 in) long, shiny, and often clustered in spur shoots; apex is acuminate; base is obtuse or rounded; margin is entire; petiole is long and exudes white sap when cut; autumn color is yellow. Twigs are moderately stout, green to orange-brown, and glabrous, with warty lenticels, a stout spine at the node, and moundlike spur shoots; leaf scar is nearly round with numerous bundle scars. A true terminal bud is lacking; lateral bud is small, round, and sunken. Flowers are dioecious; staminate flowers are in yellow-green racemes; pistillate flowers are in round hairy heads; flowers bloom in late spring after the leaves. Fruit is a collection of drupes in a large (up to 15 cm [5.9 in] wide), round, green, juicy, sticky, brainlike ball that matures in early fall. Fruit is reported to be toxic. Bark is gray-brown to orange-brown with thick interlacing ridges and orange inner bark. The growth form is a shrub often forming thickets or short tree up to 9 m (30 ft) in height with a scraggly crown.

HABITAT A range of sites, including roadsides, abandoned pastures, and fence rows.

From left to right:

Osage-orange leaves.

Osage-orange pistillate flowers.

Osage-orange staminate flowers.

From left to right:

Osage-orange fruit.

Osage-orange bark.

NOTES The original range of Osage-orange included Texas, Oklahoma, and Arkansas, mostly in bottomlands. Osage-orange was planted extensively for hedges and wind breaks and has naturalized throughout the United States. The curious fruit of Osage-orange, and other large-fruited trees such as honeylocust and Kentucky coffeetree, may have been tailored for dispersal by Pleistocene megafauna such as mastodons and giant sloths (Gill 2014). The fruit is reported to be a cockroach repellent. The wood is yellow to bright orange; very hard, heavy, and durable; and warps when drying. Native Americans (including Osage Native Americans) prized the wood for bows. A yellow dye was made from the bark during World War I. The seeds are occasionally used by squirrels and northern bobwhite. The thickets provide nesting substrate for birds and cover for other wildlife.

Maclura refers to the geologist W. Maclure; *pomifera* means "apple bearing." The hyphen in the common name indicates it is not a true orange.

Red Mulberry

Morus rubra L.

COMMON NAMES red mulberry

QUICK GUIDE Leaves alternate, simple, ovate, un-lobed or lobed, often scabrous, apex acuminate, margin serrate, petiole with white sap; fruit red-purple, juicy drupes; bark gray-brown and scaly.

DESCRIPTION Leaves are alternate, simple, deciduous, ovate or orbicular, 10–25 cm (3.9–9.8 in) long, and unlobed or with up to three lobes; apex is acuminate; base is rounded or cordate; margin is serrate; upper surface is often scabrous, with prominent veins; petiole exudes milky sap when cut; underside has hair; autumn color is yellow. Twigs zigzag and are moderately stout, yellow-brown to red-brown, black dotted, and mostly glabrous, with lenticels; leaf scar is round and sunken, with numerous bundle scars. A true terminal bud is lacking; lateral bud is ovoid and about 6 mm (0.2 in) long; scales are glabrous and maroon or green-brown, with dark margins. Flowers are usually dioecious; staminate flowers are in long yellow-green catkins; pistillate flowers are in shorter catkins; flowers bloom in the spring. Fruit is a drupe, red-purple, sweet, juicy; drupes are in clusters up to 3 cm (1.2 in) long; fruit matures in the summer. Bark is gray to red-brown

Clockwise from upper left:

Red mulberry leaf.

Red mulberry leaves with lobing.

Red mulberry staminate flowers.

Red mulberry pistillate flowers.

Red mulberry fruit.

Red mulberry
bark.

with raised ridges on small trees; on large trees the bark is gray-brown with scaly or loose ridges. The growth form is an understory tree up to 21 m (70 ft) in height and 1 m (3 ft) in diameter.

HABITAT Moist soils and in a variety of sites, including coves, stream edges, bottomlands, pastures, and forest edges.

NOTES Red mulberry is usually found growing with a wide variety of other tree species. The wood is yellow to orange-brown, heavy, and hard, and is of limited commercial use. Historical uses include boat building and fiber for making rope and coarse cloth by Native Americans. The sweet fruit is used in jams and wines, and is relished by a wide variety of songbirds and game birds, including northern bobwhite and wild turkey, foxes, squirrels, opossum, and raccoon. The foliage while succulent is browsed by white-tailed deer, and rabbits eat the bark.

Morus is Latin for "mulberry tree"; *rubra* means "red," referring to the fruit.

SIMILAR SPECIES White Mulberry (*Morus alba* L.) is originally from China and has naturalized in open areas. It is similar to red mulberry and identified by leaves with a mostly glabrous upper surface and underside; fruit that can be red, white, or purple; and bark that is ridged rather than scaly. **Paper-mulberry** (*Broussonetia papyrifera* [L.] Vent.), also originally from Asia, has naturalized in open areas and is identified by alternate, opposite, or whorled leaves, and dense long hairs on leaves and twigs.

Wax-Myrtle Family (*Myricaceae*)

Southern Bayberry

Morella cerifera (L.) Small

COMMON NAMES southern bayberry, wax-myrtle, candleberry

QUICK GUIDE Leaves alternate, simple, evergreen, and oblanceolate, with a spicy aroma and yellow resin dots on the upper surface and underside; fruit a small, waxy, blue-gray drupe; bark smooth and gray with corky lenticels.

DESCRIPTION Leaves are alternate, simple, evergreen, mostly oblanceo-late, and 3–10 cm (1.2–3.9 in) long, and smell spicy when crushed; apex is acute; base is cuneate; margin is entire or has several coarse teeth above the middle; upper surface and underside have yellow resin dots. Twigs are slender and green or brown, with brown pubescence and a spicy aroma when cut; leaf scar is crescent shaped. The terminal bud is minute, yellow, and scurfy. Flowers are dioecious; pistillate and sta-minate flowers are in oblong catkins in early spring. Fruit is a drupe, round, about 3 mm (0.1 in) wide, blue-gray, and waxy, and grows in

Clockwise from upper left:

Southern bayberry leaves.

Southern bayberry pistillate flowers.

Southern bayberry staminate flowers.

Southern bayberry fruit.

Southern bayberry
bark.

compact clusters along the branch; fruit matures in the fall. Bark is
gray-brown and smooth, with warty lenticels. The growth form is a
shrub or small tree with multiple trunks up to 12 m (40 ft) in height.

HABITAT A variety of sites, including flatwoods, swamp edges, old
fields, bogs, dry sandy soils, and mixed woodlands.

NOTES Southern bayberry is found growing with a wide variety of
tree species throughout Alabama and is planted as an ornamental for
the evergreen, fragrant foliage. The fruit was used in candle making,
and the leaves were used to repel insects and flavor soups. The fruit
is eaten by many songbirds, northern bobwhite, red-cockaded woodpecker, wild turkey, and foxes.

    Cerifera means "wax bearing," referring to the fruit.

SIMILAR SPECIES **Swamp candleberry** (*Morella caroliniensis* [Mill.]
Small) is a shrub or small tree found in swamps and bays of the lower
Southern Coastal Plain and has leaves lacking resin dots on the upper
surface and twigs with shaggy pubescence. **Odorless bayberry** (*Morella inodora* [Bartram] Small) is a shrub also found in wet areas of the
lower Southern Coastal Plain and has odorless leaves with an entire
leaf margin and a glabrous twig.

Olive Family (*Oleaceae*)

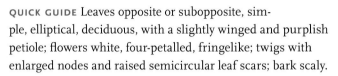

Fringe-Tree

Chionanthus virginicus L.

COMMON NAMES fringe-tree, old-man's beard, grandsy gray-beard

QUICK GUIDE Leaves opposite or subopposite, simple, elliptical, deciduous, with a slightly winged and purplish petiole; flowers white, four-petalled, fringelike; twigs with enlarged nodes and raised semicircular leaf scars; bark scaly.

DESCRIPTION Leaves are opposite or subopposite, simple, deciduous, elliptical to lanceolate or oblanceolate, and 10–20 cm (3.9–7.9 in) long; apex is acute to rounded; base is cuneate; margin is entire; petiole is somewhat winged and purplish; underside is glabrous or pubescent; autumn color is yellow. Twigs are moderately stout, red-brown to gray, glabrous or pubescent, with lenticels and enlarged nodes; leaf scar is raised, semicircular, and concave, with one linear or U-shaped bundle scar. The terminal bud is ovoid and about 6 mm (0.2 in) long; scales are overlapping, brown, keeled, and fringed with white; lateral bud is moundlike and embedded. Flowers are mostly dioecious, creamy white, and four-petalled, and bloom in fringelike clusters in early spring. Fruit is a drupe, nearly round, up to 2 cm (0.8 in) long,

Clockwise from left:

Fringe-tree leaves showing purplish base of the petiole; photo by Lisa J. Samuelson.

Fringe-tree young leaves and flowers.

Fringe-tree leaves and immature fruit.

Fringe-tree bark.

and dark blue, and matures in late summer. Bark is gray-brown and shallowly grooved on small trees; large trees have scaly ridges The growth form is up to 9 m (30 ft) in height.

HABITAT Fertile, moist soils of upland forests and along edges of streams and bogs.

NOTES Fringe-tree is found growing with a variety of tree species depending on the site. It is a very much loved southern ornamental tree because of the interesting and beautiful flowers, which bloom even on seedlings. The leaves were once used as a folk remedy for fevers and "yaws," and a decoction from the bark was used as a tonic for a variety of ailments for humans and horses. The fruit is eaten by birds and mammals, including wild turkey, northern bobwhite, and white-tailed deer.

Chionanthus is from Greek meaning "snow flower"; *virginicus* refers to the geographic range.

White Ash

Fraxinus americana L.

COMMON NAMES white ash

QUICK GUIDE Leaves opposite, pinnately compound, usually with seven leaflets; leaflets are ovate; lateral bud sits within the crescent-shaped leaf scar; fruit a single samara with a broad wing extending to the top of the seed; bark brown-gray with interlacing ridges forming a diamond pattern, sometimes blocky.

DESCRIPTION Leaves are opposite, pinnately compound, deciduous, and up to 35 cm (13.8 in) long, with five to nine leaflets. Leaflets are elliptical to ovate and 5–13 cm (2.0–5.1 in) long; margin is entire or remotely serrate; underside is pubescent or glabrous; autumn color is yellow to dull red. Twigs are moderately stout, gray-brown to green-brown, and glabrous or pubescent, with lenticels and flattened nodes; leaf scar is crescent shaped with bundle scars forming a U shape. The terminal bud is round, up to 8 mm (0.3 in) long, and dark chocolate brown, with suedelike scales; lateral bud is smaller and sits within the leaf scar. Flowers are dioecious; staminate flowers are red-green, short-stalked,

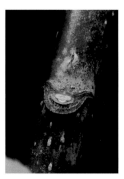

Clockwise from upper left:

White ash leaf.

White ash leaf scar.

White ash fruit.

White ash pistillate flowers.

White ash bark.

and in dense, compact clusters; pistillate flowers are vase shaped and long-stalked, with an elongated style; flowers bloom before the leaves. Fruit is a single samara, elliptical to oblong and up to 5 cm (2 in) long; wing is fairly broad and extends to the top of the seed; fruit matures in the fall. Bark is brown-gray with interlacing ridges forming a diamond pattern; large trees are deeply furrowed and sometimes blocky. The growth form is up to 30 m (100 ft) in height and 1 m (3 ft) in diameter.

HABITAT Moist, fertile soils of uplands, bottomlands, and edges of streams and rivers.

NOTES Forest associates of white ash include red maple, sugar maple, American beech, tulip-polar, eastern white pine, white oak, northern red oak, black cherry, eastern hemlock, and American elm. The emerald ash borer, a wood-boring insect from Asia that attacks all species of ash, has recently been reported in Alabama. The wood is white to light brown, heavy, hard, stiff, and straight-grained, and is used for handles, veneer, furniture, crates, pallets, and sports equipment such as Louisville Slugger baseball bats. White ash is planted as a shade tree. The seed is eaten by a variety of birds and small mammals. White-tailed deer will lightly browse the foliage, and beaver eat the bark.

Fraxinus is Latin for "ash tree"; *americana* refers to the New World.

Green Ash

Fraxinus pennsylvanica Marshall

COMMON NAMES green ash, swamp ash, water ash

QUICK GUIDE Leaves opposite, pinnately compound, usually with seven leaflets; leaflets elliptical to lanceolate; lateral bud sits above the shield-shaped leaf scar; fruit a single samara with the wing extending halfway down the narrow, elongated seed; bark brown-gray with interlacing ridges forming a diamond pattern.

DESCRIPTION Leaves are pinnately compound, opposite, deciduous, and up to 30 cm (11.8 in) long, with seven to nine leaflets. Leaflets are elliptical to lanceolate and 5–15 cm (2.0–5.9 in) long; margin is entire or remotely serrate; underside is glabrous or pubescent; autumn color is yellow. Twigs are moderately stout, gray-brown, and glabrous or pubescent, with lenticels and flattened nodes; leaf scar is shield shaped with bundle scars forming a U shape. The terminal bud is round, up to 8 mm (0.3 in) long, and dark chocolate brown, with suedelike scales; lateral bud is smaller and sits above the leaf scar. Flowers are dioecious; staminate flowers are red-green, short-stalked, and in dense compact clusters; pistillate flowers are vase shaped and long-stalked, with an elongated style; flowers appear before the leaves. Fruit is a samara, single, spatulate to oblanceolate, and up to 5 cm (2 in) long; wing narrows at the seed and extends about halfway down the seed; fruit matures in the fall. Bark is brown-gray with interlacing ridges forming a diamond pattern; large trees are deeply furrowed. The growth form is up to 36 m (120 ft) in height and 1 m (3 ft) in diameter.

HABITAT Moist to wet soils, such as moist uplands, periodically inundated bottomlands, and edges of streams and rivers.

From left to right:

Green ash leaflets.

Green ash staminate flowers.

Green ash fruit.

Green ash bark.

NOTES Green ash is associated with many species, including silver maple, boxelder, red maple, water hickory, sugarberry, white ash, sweetgum, eastern cottonwood, overcup oak, swamp chestnut oak, water oak, cherrybark oak, willow oak, northern red oak, black willow, basswood, and American elm. The emerald ash borer, a wood-boring insect from Asia that attacks all species of ash, has recently been reported in Alabama. The wood is white to light brown, heavy, hard, stiff, and straight-grained. Wood uses are similar to white ash but the quality is not as high. Green ash is planted as a shade tree. The seed is eaten by a variety of birds and small mammals. White-tailed deer will lightly browse the foliage, and beaver eat the bark.

Fraxinus is Latin for "ash tree"; *pennsylvanica* refers to the geographic range.

SIMILAR SPECIES Other ashes that can be found in Alabama include: **Carolina ash** (*Fraxinus caroliniana* Mill.), also known as water or swamp ash. Carolina ash is found on wet sites in the Southern Coastal Plain and identified by a lateral bud sitting mostly above the leaf scar; a samara that is elliptical to diamond shaped, often three winged, and up to 5 cm (2 cm) long, with the wing surrounding the seed; and by bark with scaly ridges. **Pumpkin ash** (*Fraxinus profunda* [Bush] Bush) may be found in the extreme southern portion of the state in swamps and bottoms. Pumpkin ash has seven to nine leaflets with white hair on the underside, densely pubescent twigs, and a large samara (the largest of the ashes, up to 8 cm [3.1 in] long) with the wing extending narrowly to the bottom of the seed. **Blue ash** (*Fraxinus quadrangulata* Michx.) is an occasional tree on drier upland sites, and it usually has nine leaflets, twigs that are four angled and winged (see appendix for winter twig identification photo, page 318), a samara that is oblong to obovate and up to 5 cm (2 in) long, and bark with scaly ridges and blue inner bark.

Devilwood

Osmanthus americanus (L.)
Benth. & Hook. f. ex A. Gray

COMMON NAMES devilwood, wild-olive, fragrant-olive

QUICK GUIDE Leaves opposite, simple, evergreen, elliptical, and leathery; flowers fragrant, white; fruit an olivelike dark blue drupe; bark brown-gray to black, small trees smooth and warty.

DESCRIPTION Leaves are opposite, simple, evergreen, elliptical to lanceolate or obovate, 6–15 cm (2.4–5.9 in) long, and leathery; apex is acute; base is cuneate; margin is entire and sometimes revolute; petiole is stout; underside is glabrous. Twigs are slender, green to graybrown, and smooth or scaly, with lenticels; leaf scar is widely crescent shaped and raised, with one crescent-shaped bundle scar. The terminal bud is acute and about 1 cm (0.4 in) long, with two brown valvate scales. Flowers are perfect and imperfect, small, white, and fragrant, and grow in clusters in early spring; immature flowers are prominent in the fall and winter. Fruit is a drupe, olivelike, round, up to 2 cm (0.8 in) long, and dark blue; fruit matures in the fall. Bark

Clockwise from upper left:

Devilwood leaves.

Devilwood flowers.

Devilwood leaves and fruit.

Bark of young devilwood.

is brown-gray to black, often mottled, and smooth, with lenticels or warts; large trees are darker and rougher. The growth form is a small tree up to 9 m (30 ft) in height and can form thickets.

HABITAT Near streams and other water edges; on bluffs and sandy ridges; and in hammocks and dune forests of the Coastal Plain.

NOTES Devilwood is associated with a variety of tree and shrub species. The wood is hard and difficult to work (hence the name). The fragrant flowers and evergreen foliage make this species an attractive specimen for landscaping.

Osmanthus is from Greek for "fragrant flower"; *americanus* refers to the New World.

Sycamore Family (*Platanaceae*)

Sycamore

Platanus occidentalis L.

COMMON NAMES sycamore, American sycamore, American planetree, buttonwood, buttonball-tree

QUICK GUIDE Leaves alternate, simple, large, fan shaped, with three to five lobes, margin coarsely toothed; bud concealed beneath the petiole, conical with one smooth scale; fruit a long-stalked ball of achenes; bark brown and scaly, exposing smooth white bark.

DESCRIPTION Leaves are alternate, simple, deciduous, fan shaped, and 13–20 cm (5.1–7.9 in) wide, with three to five lobes; sinuses are shallow; apex is acuminate; base is truncate; margin is wavy and coarsely toothed; petiole is stout and hollow at the base and hides the lateral bud; underside is pubescent; autumn color is dull yellow. Twigs zigzag and are moderately stout, orange or yellow-brown, and glabrous, with stipule scars; leaf scar is thin and nearly encircles the bud. A true terminal bud is lacking; lateral bud is divergent, about 8 mm (0.3 in) long, conical, and caplike, with one green-maroon glabrous scale, and concealed beneath the petiole when the leaves are present. Flowers are imperfect and in long-stalked, round heads from leaf axils; pistillate flowers are red and on longer stalks; staminate flowers are smaller, yellow-green, and on shorter stalks; flowers appear in the spring with the leaves. Fruit is an achene; many achenes are in a ball-like cluster that is up to 3 cm (1.2 in) wide and long stalked; fruit matures in the fall. Bark is brown, flaky, and scaly, revealing smooth green-white bark;

From left to right:

Sycamore leaves and stipules.

Sycamore pistillate flowers.

Sycamore fruit.

From left to right:

Sycamore with exposed pale bark.

Sycamore scaly bark.

upper trunk is smooth and white. The growth form is a very large tree that grows up to 42 m (140 ft) in height and 3 m (10 ft) in diameter.

HABITAT Banks of streams and rivers and in bottomlands.

NOTES Sycamore is found with boxelder, red maple, silver maple, river birch, sugarberry, hackberry, green ash, sweetgum, eastern cottonwood, and black willow. The wood is blond to light brown, close-grained, coarse, and moderately hard, and is used for veneer, pulpwood, inexpensive furniture, paneling, trim, butcher blocks, pallets, and boxes. The bark and leaves make for an interesting landscape tree, but this tree requires moist soils and plenty of space. The core of the fruit was used for buttons. The fruit is of minimal wildlife value, although some finches will eat the seeds. It is used as a den tree by wildlife.

Platanus is Latin for "plane-tree," referring to the broad leaves; *occidentalis* refers to the Western Hemisphere.

Buckthorn Family (*Rhamnaceae*)

Carolina Buckthorn

Frangula caroliniana (Walter) A. Gray

COMMON NAMES Carolina buckthorn, Indian cherry, yellow buckthorn

QUICK GUIDE Leaves alternate, simple, elliptical, shiny, with prominent parallel venation, margin obscurely serrate; buds naked and wooly; fruit a red to black drupe; bark gray, mottled, and mostly smooth.

DESCRIPTION Leaves are alternate, simple, deciduous, oblong to elliptical, 5–15 cm (2.0–5.9 in) long, and shiny green or yellow-green; apex is acute; base is rounded or tapered; margin is obscurely serrate; petiole is possibly red; upperside has prominent parallel venation and the veins curve at the margin; autumn color is yellow, orange, or red. Twigs are slender, red-brown, and pubescent; leaf scar is elliptical with three bundle scars. The terminal bud is about 6 mm (0.2 in) long and naked, with wooly pale hair. Flowers are perfect, greenish or yellow-white, small, and bell shaped, with five petals, and bloom in small clusters in late spring or early summer. Fruit is a drupe, round, up to 8 mm (0.3 in) wide, and red or black; fruit matures in late summer or fall. Bark is gray, mottled, and smooth, and becomes shallowly ridged on large trees. The growth form is a shrub or tree up to 12 m (40 ft) in height.

From left to right:

Carolina buckthorn leaves and fruit.

Carolina buckthorn fruit and twig.

Carolina buck-
thorn bark.

HABITAT A variety of sites such as moist woods, mesic slopes, upland mixed woods, sandy soils, stream banks, creek bottoms, glades, fence-rows, and on chalky soils.

NOTES The bark of Carolina buckthorn was used in making a yellow dye, and charcoal from the wood was used in making gunpowder. Carolina buckthorn is sometimes planted as an ornamental tree or hedge for the attractive foliage and red fruit. The fruit is eaten by birds.

Frangula is after *Rhamnus frangula* (alder buckthorn); *caroliniana* refers to the geographic range.

Rose Family (*Rosaceae*)

Downy Serviceberry

Amelanchier arborea (Michx. f.) Fern.

COMMON NAMES downy serviceberry, juneberry, shadbush

QUICK GUIDE Leaves alternate, simple, ovate, margin serrate, underside usually pubescent, petiole pubescent; flowers in white racemes before or with the appearance of the leaves; buds long-pointed; bark with vertical stripes or cracks.

DESCRIPTION Leaves are alternate, simple, deciduous, ovate to oval, and 4–10 cm (1.6–3.9 in) long; apex is acute; base is rounded to cordate; margin is serrate; petiole is pubescent; underside is downy pubescent at unfolding; autumn color is yellow to red. Twigs are slender, red-brown, and glabrous or pubescent, with white lenticels; leaf scar is crescent shaped with three bundle scars. The terminal bud is acute, long-pointed, and about 1 cm (0.4 in) long; five scales are green-red, overlapping, and glabrous or with white pubescence on the margins; lateral bud is appressed to the twig. Flowers are perfect and in short-stalked, white racemes, and bloom in early spring before or with the leaves. Fruit is a pome, nearly round, up to 1 cm (0.4 in) wide, and red to purple, and matures in early summer. Bark is green to gray-brown and smooth, with vertical stripes or streaks sometimes twisting around the tree; bark of large trees is darker, shallowly grooved, and cracked. The growth form is a shrub or tree up to 15 m (50 ft) in height.

From left to right:

Downy serviceberry leaf.

Downy serviceberry flowers.

Downy serviceberry fruit.

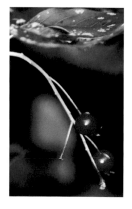

From left to right:

Downy service-berry bark with stripes.

Downy service-berry bark with cracks.

HABITAT Well-drained soils and in the understory of a variety of sites, including moist woods, dry limestone soils, and oak-pine forests.

NOTES Downy serviceberry is an attractive ornamental tree for its early spring flowers, interesting bark, autumn foliage, and tolerance of a range of conditions. The wood is red-brown and hard and is used for tool handles. Serviceberries are considered valuable wildlife foods. The fruit is eaten by a variety of songbirds and game birds, white-tailed deer, raccoon, foxes, chipmunk, bear, squirrels, and rabbits. White-tailed deer browse the foliage and twigs, and beaver eat the bark.

Amelanchier is a French name; *arborea* means "treelike."

SIMILAR SPECIES **Allegheny serviceberry** (*Amelanchier laevis* Wiegand) is similar but an uncommon tree in Alabama and is distinguished by leaves that are bronze-purple and glabrous when unfolding and a blue-black mature fruit.

Hawthorns

Crataegus spp.

COMMON NAMES hawthorns, haws

QUICK GUIDE Leaves with lobed, serrate, or dentate margins; twigs with thorns; flowers white or pink, with five petals; fruit a round pome; bark smooth on small trees, scaly or peeling on large trees.

DESCRIPTION A genus including many often-confusing species, so only basic characteristics are presented here. Leaves are alternate, simple, and deciduous with a margin that is often lobed, serrate, or dentate. Twigs often have thorns; leaf scar is crescent shaped with three bundle scars. Flowers are perfect, showy, white or pink, with five petals, and bloom in the spring. Fruit is a pome, round, 6–15 mm (0.2–0.6 in) wide, and usually matures in the fall. Bark is green to

Clockwise from upper left:

Hawthorn leaves.

Hawthorn fruit and leaves.

Hawthorn flowers and leaves.

Hawthorn bark.

maroon, brown, or black, mottled, and smooth on small trees, and is scaly or peeling on large trees. The growth form is a shrub or small tree to 6 m (20 ft), some species have a fluted trunk.

HABITAT A variety of sites.

NOTES The wood of *Crataegus* species is hard and strong and is used for small specialty items such as tool handles. Many cultivars are available for landscaping. The fruit is used in jellies and preserves. The fruit is eaten by white-tailed deer, black bear, foxes, coyotes, songbirds and game birds, and small mammals. Thickets provide excellent nesting cover for a variety of songbirds.

Crataegus is derived from a Greek work for "strength," referring to the hard wood.

Southern Crabapple

Malus angustifolia (Aiton) Michx.

COMMON NAMES southern crabapple, narrow-leaved crabapple

QUICK GUIDE Leaves alternate, simple, elliptical, margin finely serrate to entire, base sometimes lobed; twigs often with thorns; flowers in showy, fragrant, white-pink racemes with the leaves; bark scaly and gray-brown.

DESCRIPTION Leaves are alternate, simple, deciduous, elliptical to ovate or oblong, and 2.5–8 cm (1.0–3.1 in) long; margin ranges from finely serrate to crenate or entire and is sometimes lobed on spur shoots; petiole is pubescent; underside is pubescent only when young; autumn color is yellow. Twigs are red-brown and pubescent; leaf scar is crescent shaped with three bundle scars; spur shoots often end in thorns. Terminal bud is small (about 3 mm [0.1 in] long), with brown overlapping and pubescent scales. Flowers are perfect, white-pink, and fragrant, with five petals, and bloom in terminal racemes with the leaves. Fruit is a pome, round, about 3 cm (1.2 in) wide, yellow-green to red and sour, and matures in the fall. Bark is gray-brown to red-brown and scaly. The growth form is a shrub or small tree up to 9 m (30 ft) in height and can form thickets.

HABITAT A variety of sites, including moist woodlands and along fencerows and stream edges.

NOTES A wide selection of southern crabapple cultivars are available for landscaping. The fruit is used in jellies and preserves. The fruit is

From left to right:

Southern crabapple leaves.

Southern crabapple fruit.

Southern crabapple flower.

Southern crab-
apple bark.

eaten by white-tailed deer, wild turkey, various songbirds, and small mammals. Thickets form excellent nesting and escape cover for a variety of songbirds and game birds.

Malus is Latin for "apple tree"; *angustifolia* means "narrow leaf."

Chickasaw Plum

Prunus angustifolia Marshall

COMMON NAMES Chickasaw plum, wild plum

QUICK GUIDE Leaves alternate, simple, lanceolate, often folding upward, margin with gland-tipped teeth, petiole red; twigs shiny or glaucous, with thorns.

DESCRIPTION Leaves are alternate, simple, deciduous, elliptical to lanceolate, 3–8 cm (1.2–3.1 in) long, and often folding upward; apex is acuminate; base is acute; margin is serrate and teeth are tipped with yellow or red glands; petiole is red with one or two glands near the blade; autumn color is yellow. Twigs are slender, red-brown, and shiny or glaucous, with lenticels and thorns, and may end in a thorn; leaf scar is small and semicircular, with three bundle scars. A true terminal bud is lacking; lateral bud is acute and about 5 mm (0.2 in) long, with red-brown, overlapping, loose scales. Flowers are perfect and bloom in showy white umbels before the leaves. Fruit is a drupe, round, 1.0–1.5 cm (0.4–0.6 in) wide, yellow or red, and sweet, and

Clockwise from upper left:

Chickasaw plum leaves.

Chickasaw plum fruit.

Chickasaw plum flowers.

From left to right:

Young Chickasaw plum with thorns.

Chickasaw plum bark.

matures in late summer. Bark is red-brown and shiny with horizontal lenticels on small trees; large trees are scalier. The growth form is a shrub or small tree up to 6 m (20 ft) and often forms thickets as a result of root sprouting.

HABITAT Fencerows, old fields, disturbed areas.

NOTES Chickasaw plum is a small tree most often found growing in the open in small thickets. The fruit is used in jellies and preserves. White-tailed deer, black bear, foxes, raccoon, and opossum eat the fruit. Thickets form excellent nesting and escape cover for a variety of songbirds and game birds.

Prunus is Latin for "plum tree"; *angustifolia* means "narrow leaf."

SIMILAR SPECIES **Flatwoods plum** or **hog plum** (*Prunus umbellata* Elliott) is similar to Chickasaw plum but does not form thickets and is distinguished by elliptical to ovate leaves with margin teeth lacking glands, a downy leaf midrib, and smaller purple fruit. **American plum** (*Prunus americana* Marshall) is found on a variety of open sites and is distinguished by more ovate leaves with an acuminate apex and sharply serrate margin lacking glands, flowers about double in size, and a larger fruit.

Cherry Laurel

Prunus caroliniana (Mill.) Aiton

COMMON NAMES cherry laurel, Carolina laurel cherry, laurel cherry

QUICK GUIDE Leaves alternate, simple, evergreen, shiny, elliptical, margin with an occasional hooked red tooth; fruit a blue-black drupe persisting over winter; bark gray-brown and smooth.

DESCRIPTION Leaves are alternate, simple, evergreen, elliptical to oblanceolate, 5–12 cm (2.0–4.7 in) long, leathery, and shiny; apex and base are acute; margin is entire or with several hooked red teeth; petiole is red and lacks glands; underside is glabrous. Twigs are slender, red-brown, and glaucous, with lenticels; leaf scar is small, heart shaped, and raised, with three bundle scars. The terminal bud is acute and about 6 mm (0.2 in) long, with red-black overlapping scales; flower buds are plump and red-white. Flowers are perfect, yellow-white, and short-stalked, and bloom in erect racemes in the spring before the new leaves. Fruit is a drupe, nearly round, about 1.5 cm (0.6 in) wide, and blue-black; fruit matures in late summer or fall and persists over the winter. Bark is gray-brown and mostly smooth, with horizontal lenticels. The growth form is an understory tree up to 9 m (30 ft) in height.

Clockwise from left:

Cherry laurel leaves, *note the margin teeth.*

Cherry laurel flowers.

Cherry laurel fruit.

Cherry laurel bark.

HABITAT A variety of habitat types, including upland forests, near rivers and streams, along fencerows, in open disturbed areas, and in coastal forests.

NOTES The leaves and twigs of cherry laurel may be poisonous to livestock. The fruit is relished by songbirds, particularly by American robins and cedar waxwings.

Prunus is Latin for "plum tree"; *caroliniana* refers to the geographic range.

Mexican Plum

Prunus mexicana S. Watson

COMMON NAMES Mexican plum, bigtree plum, fall plum

QUICK GUIDE Leaves alternate, simple, ovate, with prominent venation and a sharply serrate margin, underside densely pubescent; twigs with and without thorns; bark with rough, scaly ridges.

DESCRIPTION Leaves are alternate, simple, deciduous, ovate to elliptical, and 5–13 cm (2.0–5.1 in) long; apex is acute or acuminate; base is acute to rounded; margin is sharply serrate; petiole is red and pubescent and may have glands at the leaf base; upperside has deeply sunken veins; underside is densely pubescent; autumn color is yellow. Twigs are stiff, pubescent when young, maroon-gray, and glaucous or glossy, with lenticels and spur branches, and may be found with and without thorns; leaf scar is semicircular with hair on the top margin and three bundle scars. The terminal bud is acute, about 6 mm (0.2 in) long, with overlapping brown-maroon scales. Flowers are perfect and bloom in showy white umbels before or with the leaves. Fruit is a drupe, 2.5–4.0 cm (1.0–1.6 in) wide, nearly round, sweet, glaucous, and red to purplish when mature in late summer. Bark is maroon to dark gray and scaly, with horizontal lenticels; bark of large trees is furrowed with rough, loose ridges. The growth form is up to 9 m (30 ft) in height and does not form thickets.

Clockwise from left:

Mexican plum leaves.

Mexican plum flowers.

Mexican plum immature fruit.

Mexican plum
bark.

HABITAT A variety of sites.

NOTES The fruit is used in jams and preserves. The flowers are vis-ited by bees, and a variety of birds and mammals eat the fruit. Mexi-can plum is planted as an ornamental for its flowers and tolerance of dry soils.

Prunus is Latin for "plum tree"; *mexicana* refers to the geographic range.

Black Cherry

Prunus serotina Ehrh.

COMMON NAMES black cherry, rum cherry, wild black cherry

QUICK GUIDE Leaves alternate, simple, and oval to oblong or lanceolate, margin serrate, midrib with tawny pubescence; petiole with one or two glands near the blade; flowers in white racemes after the leaves; fruit a drupe, purple-black, juicy; bark with black scales.

DESCRIPTION Leaves are alternate, simple, deciduous, oval to oblong or lanceolate, 5–15 cm (2.0–5.9 in) long, and shiny; apex is acuminate; base is obtuse to rounded; margin is serrate; midrib has tawny pubescence; petiole has one or two glands near the blade; autumn color is yellow. Twigs are slender, red-brown, and shiny or glaucous, with lenticels and an almondlike smell when cut; leaf scar is small and semicircular, with three bundle scars. The terminal bud is acute and about 4 mm (0.2 in) long; scales are overlapping and vary in color from brown to red-brown, red-green, or pinkish. Flowers are perfect and bloom in long white racemes in the spring with the young leaves. Fruit is a drupe, round, 0.8–1.3 cm (0.3–0.5 in) wide, and red-purple to black, and matures in the summer. Bark on small trees is red-brown, smooth, and lustrous, with horizontal lenticels; bark on large

Clockwise from left:

Black cherry leaves.

Black cherry flowers.

Black cherry fruit.

From left to right:

Scaly bark of a black cherry tree.

Bark of a large black cherry tree.

trees has gray-black scales or plates. The growth form is up to 24 m (80 ft) in height and 1 m (3 ft) in diameter.

HABITAT A variety of soils but best growth on moist, fertile soils.

NOTES Forest associates of black cherry are numerous and site dependent, and on mesic sites include red maple, sugar maple, black birch, American beech, tulip-poplar, eastern white pine, white oak, northern red oak, and eastern hemlock. The wood is red-brown, lustrous when finished, moderately heavy, moderately hard, straight-grained, and commercially very valuable, and is used for furniture, cabinets, veneer, and woodenware. The fruit was added to cordials, brandies, and rum as a flavoring. Black cherry is susceptible to fungal disease and infestation by the forest tent caterpillar. The wilted foliage may be poisonous to livestock because of the release of hydrocyanic acid. The fruit is important to wildlife because of the consistency of production and maturation in the summer when few other fruits are available. The fruit is eaten by a variety of birds, including northern bobwhite, and numerous songbirds, and is also eaten by black bear, foxes, raccoon, squirrels, opossum, and numerous small mammals.

Prunus is Latin for "plum tree"; *serotina* means "late ripening or late opening."

SIMILAR SPECIES Alabama black cherry or **Alabama chokecherry** (*Prunus alabamensis* C. Mohr) is an uncommon tree reported in the mountains and identified by oval to ovate leaves with hair on the underside, margins with gland-tipped teeth, and a glandular hairy petiole.

Rue Family (*Rutaceae*)

Hercules'-Club

Zanthoxylum clava-herculis L.

COMMON NAMES Hercules'-club, toothache-tree, southern prickly-ash

QUICK GUIDE Leaves alternate and pinnately compound; leaflets shiny and falcate, margins crenate with yellow glands between margin teeth; spines often on the rachis, petiole, and young branches; bark with sharp or corky pyramidal growths.

DESCRIPTION Leaves are alternate, pinnately compound, deciduous, and up to 38 cm (15 in) long, and have 7–9 (possibly 5–19) leaflets. Leaflets are lanceolate to ovate or falcate, 2.5–7.0 cm (1–10 in) long, and shiny; apex is acute or acuminate; base is inequilateral or rounded; margin is crenate with yellow glands between the teeth; petiole is notably red or purplish; underside is glandular; rachis is with and without spines; leaves have a rank odor when crushed; autumn color is yellow. Twigs are stout, green-brown, and glabrous, with scattered sharp spines; leaf scar is shield shaped with three bundle scars. The terminal bud is small and round; scales are pubescent and green-brown, with red-brown dots. Flowers are perfect or imperfect, and often dioecious, with yellow-green petals, and bloom in small terminal clusters in the spring or early summer. Fruit is a follicle, nearly round, and about 6 mm (0.2 in) wide, with a shiny black seed, and matures in late summer. Bark is brown-gray and smooth, with lenticels and sharp prickles on small trees; large trees have dull, corky, and pyramidal growths

From left to right:

Hercules'-club leaf.

Hercules'-club flowers.

Hercules'-club fruit.

Hercules'-club
bark.

(like the mythical club of Hercules). The growth form is up to 12 m (40 ft) in height and 46 cm (18.1 in) in diameter.

HABITAT Sandy or limestone soils, including sand dunes, coastal forests, wood margins, and along fence rows.

NOTES The bark of Hercules'-club is chewed as a folk-remedy to numb toothaches. Birds will eat the fruit.

Zanthoxylum means "golden wood"; *clava* means "a club or knotty wood"; *herculis* refers to Hercules' club.

Willow or Poplar Family (*Salicaceae*)

Eastern Cottonwood

Populus deltoides Bartram ex Marshall

COMMON NAMES eastern cottonwood, southern cottonwood, eastern poplar

QUICK GUIDE Leaves alternate, simple, triangular, large, apex acuminate, margin with coarse rounded teeth, petiole long and flattened; twig and terminal bud stout; bark ash gray and deeply grooved.

DESCRIPTION Leaves are alternate, simple, deciduous, and triangular; blade is 8–18 cm (3.1–7.1 in) long; apex is acute or acuminate; base is truncate; margin has coarse rounded teeth; petiole is flattened and long (10 cm [3.9 in]), with glands near the blade, and shakes in the wind; underside is glabrous; autumn color is yellow. Twigs are stout, yellow-brown, glabrous, and angled; leaf scar is heart shaped with three bundle scars. The terminal bud is stout, acute, angled, and up to 2 cm (0.8 cm) long; scales are overlapping, yellow-green, shiny, and sticky. Flowers are dioecious; staminate and pistillate flowers appear in catkins before the leaves. Fruit is a capsule, conical, two- to four-valved, and 6–12 mm (0.2–0.5 in) long, and releases cottonlike seeds

Clockwise from upper left:

Eastern cottonwood leaves.

Eastern cottonwood pistillate flowers.

Eastern cottonwood fruit with cottonlike seeds.

Eastern cottonwood staminate flowers.

Eastern cotton-
wood bark.

in late spring. Bark is yellow-green to light gray-brown and smooth
or shallowly grooved on small trees; large trees are light brown to ash
gray and deeply grooved with thick ridges. The growth form is up to
30 m (100 ft) in height and 1 m (4 ft) in diameter.

HABITAT Bottomlands, and edges of rivers and streams.

NOTES Eastern cottonwood is a fast-growing tree found with boxelder,
red maple, silver maple, river birch, numerous ashes, sweetgum, sycamore, willow oak, black willow, and American elm. The wood is white
to gray, light, and soft, and is used for pulpwood, veneer, crates, and
fuel. Because of very early rapid growth, it is used for windbreaks and
waste site reclamation and in short-rotation woody-crop plantations.
The leaves are browsed somewhat by deer and rabbits, and catkins
and seeds are used by songbirds. The bark is eaten by beaver.

Populus is Latin for "poplar tree"; *deltoides* refers to the Greek letter
delta and the triangular leaves.

SIMILAR SPECIES **Swamp cottonwood** (*Populus heterophylla* L.) is found
primarily in the Southern Coastal Plain in wet clay bottoms and
swamps and is distinguished by ovate leaves with smaller teeth on the
margin, a mostly acute apex, and a more rounded or cordate leaf base.

Black Willow

Salix nigra Marshall

COMMON NAMES black willow, swamp willow

QUICK GUIDE Leaves alternate, lanceolate, margin with small glandular teeth; bud with one scale; bark gray-brown to red-brown with scaly ridges or loose plates; branches slender, brittle, and yellow-orange.

DESCRIPTION Leaves are alternate, simple, deciduous, lanceolate, 8–15 cm (3.1–5.9 in) long, and nearly sessile; apex is acuminate; base is acute to rounded; margin is finely serrate and teeth are tipped with red glands; underside is mostly glabrous; autumn color is yellow. Twigs are slender, brittle, yellow-orange to maroon, and pubescent or glabrous, with stipule scars; leaf scar is V-shaped with three bundle scars. A true terminal bud is lacking; lateral bud is divergent, small (2–5 mm [0.1–0.2 in] long), and conical, with one yellow-red glabrous scale. Flowers are dioecious; staminate and pistillate flowers appear in catkins with the leaves. Fruit is a capsule, conical, two-valved, and up to 6 mm (0.2 in) long, and releases cottonlike seeds in late spring or early summer. Bark is gray-brown to dark red-brown and scaly or with

Clockwise from left:

Black willow leaves.

Black willow staminate flowers.

Black willow fruit with cottonlike seeds.

Black willow pistillate flowers

From left to right:

Black willow bark.

Loose plates of black willow bark.

long, loose plates; inner bark is chocolate brown. The growth form is up to 15 m (50 ft) in height but trees can be larger on good sites.

HABITAT Floodplains and margins of streams, rivers, and swamps.

NOTES Black willow is found with many species, including boxelder, red maple, silver maple, river birch, numerous ashes, black walnut, sweetbay magnolia, swamp and water tupelo, sycamore, bottomland oaks, and eastern cottonwood. The wood is white to red-brown, light, and soft, and is used for pulpwood, shipping boxes, baskets, caskets, furniture, and crates. In the past, charcoal from the wood was used in making gun powder. The inner bark was used to make aspirin (modern aspirin is synthetically made). Black willow is planted for soil stabilization along stream banks. The leaves are browsed by deer, and the bark and twigs are eaten by beaver. Bees will visit the flowers.

Salix is Latin for "willow"; *nigra* means "dark," referring to the bark.

Soapberry Family (*Sapindaceae*)

Yellow Buckeye

Aesculus flava Sol.

COMMON NAMES yellow buckeye, big buckeye, sweet buckeye

QUICK GUIDE Leaves opposite, palmately compound, with five leaflets; buds stout, buff colored, glabrous; flowers yellow with petals longer than stamens; fruit a globose, smooth capsule with large maroon seeds; bark smooth, then loosely plated with "bull's-eye" lines.

DESCRIPTION Leaves are opposite, palmately compound, deciduous, and up to 35 cm (13.8 in) long, with five leaflets. Leaflets are obovate to ovate or elliptical, and 10–20 cm (3.9–7.9 in) long; margin is serrate; underside is glabrous; petiole is long (up to 15 cm [5.9 in]); autumn color is yellow. Twigs are stout, yellow-brown, and glabrous, with lenticels, and do not have an unpleasant odor when cut; leaf scar is large and shield shaped with bundle scars forming a V shape. Terminal bud is stout, up to 2 cm (0.8 in) long, and acute; scales are overlapping, buff colored, and glabrous. Flowers are usually perfect; petals are usually longer than the stamens; flowers bloom in erect yellow panicles in the spring after the leaves. Fruit is a capsule, round to pear shaped, 3–4 cm (1.2–1.6 in) wide, brown, leathery, and mostly smooth, and contains one to three shiny maroon seeds with a pale spot (buck's eye). The fruit matures in the fall. Bark is gray-brown and smooth

Clockwise from left:

Yellow buckeye leaf.

Yellow buckeye flowers; courtesy of John Seiler.

Yellow buckeye fruit.

Yellow buckeye seeds showing "buck's eye".

Yellow buckeye bark.

with lenticels on small trees; large trees have scaly and loosely plated bark with "bull's-eye" lines. The growth form is up to 26 m (85 ft) in height and 1.5 m (5 ft) in diameter.

HABITAT Moist, fertile soils of cool mountain slopes, coves, river bottoms, and stream edges.

NOTES Yellow buckeye is usually found only in northern Alabama. Forest associates of yellow buckeye include sugar maple, black birch, white ash, yellow-poplar, eastern white pine, white oak, northern red oak, basswood, and eastern hemlock. The wood is yellow-white to gray, light, and very soft, and is used for pulpwood, novelty items, boxes, and crates. The wood was used for artificial limbs, and a paste made from the fruit was used in bookbinding. The smooth seed is carried as a good luck piece. Young foliage and seeds are poisonous to humans and livestock. Gray and fox squirrels and feral hogs will occasionally eat the seeds.

Aesculus is a Latin name for "a mast-bearing tree"; *flava* means "golden yellow," referring to the flowers.

SIMILAR SPECIES **Red buckeye** (*Aesculus pavia* L.) is a shrub or small tree found on mesic sites throughout Alabama and distinguished by bright red flowers and a smooth capsule. **Painted buckeye** (*Aesculus sylvatica* Bartram) is a shrub found in northern Alabama and has yellow-green-red flowers. **Bottlebrush buckeye** (*Aesculus parviflora* Walter) is found mainly in central Alabama and has small white flowers with greatly exserted stamens in erect clusters, creating a bottlebrush appearance.

Ohio Buckeye

Aesculus glabra Willd.

COMMON NAMES Ohio buckeye, stinking buckeye, fetid buckeye

QUICK GUIDE Leaves opposite, palmately compound, with five leaflets; leaves and twigs smelling unpleasant when crushed; buds with keeled scales; flowers yellow with exserted stamens; fruit a prickly or spiny capsule; bark with scaly plates.

DESCRIPTION Leaves are opposite, palmately compound, deciduous, and up to 30 cm (11.8 in) long, with five leaflets, and emit an unpleasant odor when crushed. Leaflets are obovate to elliptical and 8–15 cm (3.1–5.9 in) long; margin is serrate; petiole is long (up to 15 cm [5.9 in]); underside is glabrous; autumn color is yellow. Twigs are stout, yellow-brown to maroon, and glabrous or pubescent, with lenticels; leaf scar is large and shield shaped with bundle scars forming a V shape. Terminal bud is stout, up to 2 cm (0.8 in) long, and acute; scales are overlapping, maroon-brown, and keeled. Flowers are usually perfect and unpleasant smelling; petals are shorter than the stamens; flowers bloom in erect yellow panicles in the spring with the leaves. Fruit is a capsule, up to 5 cm (2 in) wide, round to pear shaped, yellow-brown, and leathery, with short spines, and contains one to three shiny maroon seeds with a pale spot (buck's eye). The fruit matures in the fall. Bark is gray-brown with corky ridges on small trees; large trees have scaly plates. The growth form is up to 20 m (66 ft) in height and 61 cm (24 in) in diameter.

HABITAT Best development is on rich, moist soils but can be found on drier sites.

From left to right:

Ohio buckeye leaf.

Ohio buckeye flowers; courtesy of John Seiler.

Ohio buckeye spiny fruit.

Ohio buckeye bark.

NOTES Ohio buckeye is usually found in the north and west part of the state. Forest associates include sugar maple, American beech, white ash, black walnut, black cherry, basswood, and American elm. The wood is yellow-white to gray, light, and very soft, and is used for pulpwood, novelty items, boxes, and crates. The wood was used for artificial limbs. The young foliage and fruit is poisonous to humans and livestock. Squirrels and feral hogs will occasionally eat the seeds. Ohio buckeye has limited value as a landscape tree because of disease problems.

Aesculus is Latin for "a mast-bearing tree"; *glabra* means "smooth," referring to the flower pedicels, which lack the glandular hairs seen on flower pedicels of yellow buckeye.

Florida Maple

Acer floridanum (Chapm.) Pax

COMMON NAMES Florida maple, southern sugar maple

QUICK GUIDE Leaves opposite, simple, palmately three- to five-lobed with some showing blunt apices, underside white and pubescent; fruit a double samara with divergent wings; bark gray-black with scaly plates on large trees.

DESCRIPTION Leaves are opposite, simple, deciduous, and nearly round, with three to five palmate lobes; blade is 3–9 cm (1.2–3.5 in) long; apices are blunt to acute; margin is not serrate; underside is white and usually pubescent; autumn color is red, orange, or yellow. Twigs are red-brown and pubescent, with lenticels; leaf scar is V-shaped with three bundle scars. The terminal bud is ovoid and about 3 mm (0.1 in) long; scales are overlapping, red-brown, and pubescent. Flowers are polygamous; staminate and pistillate flowers grow in long, dangling, green-yellow or yellow clusters before or with the leaves. Fruit is a green to brown double samara; wings are up to 3 cm (1.2 in) long and

Clockwise from upper left:

Florida maple leaves.

Florida maple pistillate flowers.

Florida maple staminate flowers.

Florida maple fruit.

Florida maple
bark.

form a 100-degree angle; fruit matures in the summer. Bark is gray-brown and smooth on small trees; bark on large trees is gray-black with scaly plates. The growth form is usually an understory tree but sometimes grows up to 18 m (60 ft).

HABITAT Fertile, moist soils of stream banks and coves.

NOTES Florida maple is usually an understory tree found growing with red maple, American beech, green ash, sweetgum, tulip-poplar, white oak, northern red oak, and eastern hemlock. The wood is of limited economic importance but is sometimes used for pulpwood, veneer, furniture, and flooring. There are reports of this species being tapped for syrup. The foliage is a browse for white-tailed deer, and the seed is eaten by birds and small to midsize mammals. This tree makes an attractive landscape tree due to its good form, heat tolerance, and attractive fall color.

Acer is Latin for "maple tree" and refers to the hardness of the wood; *floridanum* refers to Florida or the South.

Chalk Maple

Acer leucoderme Small

COMMON NAMES chalk maple, white-bark maple

QUICK GUIDE Leaves opposite, simple, palmately three- to five-lobed with attenuated apices, underside green with or without pubescence; fruit a double samara with widely divergent wings; bark scaly with a chalky white surface on large trees.

DESCRIPTION Leaves are opposite, simple, deciduous, and nearly round, with three to five (usually three) palmate lobes; blade is 3–9 cm (1.2–3.5 in) long; apices are attenuated; margin is not serrate; underside is green and often pubescent; autumn color is red, orange, or yellow. Twigs are red-brown and glabrous, with lenticels; leaf scar is V-shaped with three bundle scars. The terminal bud is ovoid and about 3 mm (0.1 in) long; scales are overlapping, red-brown, and pubescent. Flowers are polygamous; staminate and pistillate flowers are in long, dangling, green-yellow or yellow clusters before or with the leaves. Fruit is a green to brown double samara; wings are up to 3 cm (1.2 in) long and form a 120-degree angle; fruit matures in the summer. Bark is brown-gray and smooth on small trees; on large trees bark is chalky white (usually from lichens) and flaky, and it is dark at the base. The growth form is usually an understory tree but sometimes grows up to 18 m (60 ft).

Clockwise from left:

Chalk maple leaf.

Chalk maple leaves; photo by Lisa J. Samuelson.

Chalk maple fruit.

Bark of small chalk maple tree; photo by Lisa J. Samuelson.

HABITAT Fertile, moist soils and well-drained bottomlands.

NOTES When ranges overlap, forest associates of chalk maple are similar to that of Florida maple and include red maple, American beech, green ash, sweetgum, tulip-poplar, white oak, northern red oak, and eastern hemlock. The wood is of limited economic importance but is sometimes used for pulpwood, veneer, furniture, and flooring. The foliage is a browse for white-tailed deer, and the seed is eaten by birds and small to midsize mammals.

Acer is Latin for "maple tree" and refers to the hardness of the wood; *leucoderme* means "white skin" and refers to the bark.

Boxelder

Acer negundo L.

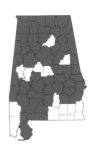

COMMON NAMES boxelder, ashleaf maple

QUICK GUIDE Leaves opposite, pinnately compound, with three to seven leaflets, margin with coarse teeth and shallow lobes; twig bright green; fruit a double samara with a flattened seed cavity; bark gray-brown and grooved.

DESCRIPTION Leaves are opposite, pinnately compound, deciduous, and up to 25 cm (9.8 in) long but can be longer, with a petiole swollen at the base and three to seven (occasionally nine) leaflets. Leaflets are ovate to elliptical and 5–10 cm (2.0–3.9 in) long; margin is coarsely serrate and often shallowly lobed; autumn color is yellow. Twigs are bright green, glabrous, and possibly glaucous, with lenticels; leaf scar is very narrowly V-shaped with three bundle scars, and opposite leaf scars meet. The terminal bud is ovoid to acute and about 5 mm (0.2 in) long; scales are overlapping, yellow-green, and pubescent; lateral bud is yellow-white and pubescent. Flowers are dioecious; staminate flowers are in long, dangling, yellow-green or sometimes red clusters on hairy stalks; pistillate flowers are in smaller drooping clusters; flowers

Clockwise from upper left:

Boxelder leaf.

Boxelder staminate flowers.

Boxelder fruit.

Boxelder pistillate flowers.

Boxelder bark.

bloom before or with the leaves. Fruit is a double samara, green to brown; wings are 2.5–4 cm (1.0–1.6 in) long and form a 90-degree angle; seed cavity is long, narrow, and flattened; fuit matures in the summer. Bark is green-gray and smooth on small trees; large trees are gray-brown and shallowly grooved. The growth form is up to 23 m (75 ft) in height and 1 m (3 ft) in diameter.

HABITAT Floodplains and on edges of streams and swamps but also found on upland sites.

NOTES Boxelder is a fast-growing, short-lived tree found with red maple, silver maple, river birch, sugarberry, ashes, sweetgum, sycamore, eastern cottonwood, bottomland oaks, black willow, and American elm. The wood is weak, brittle, and soft, and is used for fuel, crates, woodenware, and pulpwood. The seeds are eaten by birds and small to midsize mammals. Because of its fast growth, this tree is used in windbreaks and to control soil erosion. It is a poor landscape tree due to excessive seeding and sprouting, poor form, and brittle branches.

Acer is Latin for "maple tree" and refers to the hardness of the wood; *negundo* refers to similarities with *Vitex negundo*.

Red Maple

Acer rubrum L. var. *rubrum*

COMMON NAMES red maple, soft maple

QUICK GUIDE Leaves opposite, simple, palmately three- to five-lobed, sinuses shallow, margin irregularly and doubly serrate; fruit a bright red to brown double samara; bark plated and scaly on large trees.

DESCRIPTION Leaves are opposite, simple, deciduous, nearly round, and 6–19 cm (2.4–7.5 in) long, with three to five palmate lobes; apices are mostly acute; sinuses are acute to shallow; margin is irregularly and doubly serrate; underside is pale and glabrous; petiole is often red; autumn color is yellow, orange, or scarlet. Twigs are shiny and red-brown, with lenticels; leaf scar is crescent shaped with three bundle scars. The terminal bud is blunt and about 4 mm (0.2 in) long; scales are overlapping, red-brown, and mostly glabrous; flower buds are round and plump. Flowers are polygamous; staminate and pistillate flowers are in red or sometimes yellow clusters before the leaves. Fruit

Clockwise from upper left:

Red maple leaves.

Red maple pistillate flowers.

Red maple staminate flowers.

Red maple fruit.

Red maple bark.

is a bright red to brown double samara; wings are about 2 cm (0.8 in) long and form a 70-degree angle; the fruit matures in the spring. Bark is highly variable, brown-gray to white, and smooth on small trees; on large trees the bark is dark gray to brown, loosely plated or scaly. The growth form is up to 30 m (100 ft) in height and 1 m (3 ft) in diameter.

HABITAT A variety of sites, such as dry ridges, rocky uplands, stream borders, cove forests, bottomlands, and swamps.

NOTES Red maple is a species with broad ecological amplitude and is an associate of many tree species. The wood is white to gray-green, close-grained, and moderately hard. The foliage is a browse for white-tailed deer, and the seed is eaten by birds and small to midsize mammals. Honey bees favor the flowers. Red maple is a common ornamental tree because of its attractive flowers, brilliant fall color, rapid growth, and tolerance of a variety of sites. Many cultivars have been developed to enhance form and fall color.

Acer is Latin for "maple tree" and refers to the hardness of the wood; *rubrum* means "red" and describes the color of the flowers, fruit, and autumn leaves.

SIMILAR SPECIES Drummond maple (*Acer rubrum* L. var. *drummondi* [Hook. & Arn. ex Nutt.] Sarg.) is similar to red maple and is found in swamps and wetlands. It has leaves that are densely pubescent on the underside.

Silver Maple

Acer saccharinum L.

COMMON NAMES silver maple, soft maple, white maple

QUICK GUIDE Leaves opposite, simple, with five palmate lobes, margin coarsely serrate, sinuses deeply V-shaped, underside silver-white; fruit a large double samara with widely divergent wings; bark with loose, silver-gray plates.

DESCRIPTION Leaves are opposite, simple, deciduous, nearly round, and 7–20 cm (2.8–7.9 in) long, with five palmate lobes; apices are acute or acuminate; sinuses are deeply V-shaped; margin is coarsely serrate; underside is prominently silver-white; autumn color is yellow. Twigs are red-brown and shiny or glaucous, with lenticels, and emit an unpleasant odor when cut; leaf scar is crescent shaped with three bundle scars. The terminal bud is blunt and about 5 mm (0.2 in) long; scales are overlapping, red-brown, and tipped with pubescence; flower buds are round and plump. Flowers are polygamous; staminate and pistillate flowers are in dense clusters on short stalks before the leaves. Fruit is a double samara; it is large (the largest of the native maples), green to brown, and tomentose when young; wings are 4–6 cm (1.6–2.4 in) long and form a 130-degree angle, often with only one wing producing seed; fruit matures in the spring. Bark is light gray and smooth on small trees; large trees have long, loose, silver-gray plates. The growth form is up to 30 m (98 ft) in height and 1 m (3 ft) in diameter.

HABITAT Well-drained to poorly drained soils of stream banks, lake edges, and floodplains.

From left to right:

Silver maple leaf.

Silver maple flowers.

Silver maple fruit.

Silver maple bark.

NOTES Silver maple is a fast-growing tree and a prolific seeder. Forest associates include red maple, river birch, numerous ashes, sweetgum, sycamore, eastern cottonwood, overcup oak, water oak, willow oak, northern red oak, black willow, and American elm. The wood is white to gray-green and moderately hard, and is used for pulpwood, paneling, boxes, pallets, and furniture. The seed is eaten by birds and small to midsize mammals. Silver maple is planted as an ornamental tree for its fast growth, attractive leaves, and low crown, but it can be a poor choice because of weak limbs, excessive sprouting, and an invasive, shallow root system.

Acer is Latin for "maple tree" and refers to the hardness of the wood; *saccharinum* means "sugary" and refers to sap, which was occasionally boiled for sugar.

Sugar Maple

Acer saccharum Marshall

COMMON NAMES sugar maple, hard maple, sweet maple

QUICK GUIDE Leaves opposite, simple, with five palmate lobes, aspices attenuated, margin smooth, sinuses U-shaped, underside pale and glabrous; fruit a double samara with narrow wings and a swollen seed cavity; bark loose and flaky on large trees.

DESCRIPTION Leaves are opposite, simple, deciduous, nearly round, and 7–20 cm (2.8–7.9 in) long, with five palmate lobes; apices are attenuated; sinuses are U-shaped; margin is not serrate; underside is pale and glabrous; autumn color is brilliant red, orange, and yellow. Twigs are brown and shiny, with lenticels; leaf scar is V-shaped with three bundle scars. The terminal bud is acute and about 7 mm (0.3 in) long; scales are overlapping and a dark maroon-brown, with pale pubescence. Flowers are polygamous; staminate and pistillate flowers are in yellow clusters on long, drooping pubescent stalks with the leaves. Fruit is a green to brown double samara; wings are 2.5–4.0 cm (1.0–1.6 in) long and form a 50-degree angle; seed cavity is swollen; fruit matures in late summer. Bark is brown-gray and smooth or ridged on small trees; bark on large trees is flaky or scaly, with loose, curved plates. The growth form is up to 35 m (115 ft) in height and 1 m (3 ft) in diameter.

Clockwise from left:

Sugar maple leaves.

Sugar maple flowers.

Sugar maple fruit.

Sugar maple bark.

HABITAT Moist, fertile soils and cool sites.

NOTES Sugar maple is a very shade-tolerant tree found growing with red maple, yellow buckeye, black birch, American beech, Carolina silverbell, yellow-polar, cucumbertree, black cherry, eastern white pine, basswood, white oak, northern red oak, and eastern hemlock. The wood is blond, close-grained, hard, and heavy, and is prized for furniture, cabinets, flooring, and woodenware. The wood is also used for veneer, plywood, paneling, gunstocks, tool handles, musical instruments, and bowling pins. This tree is the main source of maple syrup. Sugar maple is a browse for white-tailed deer and rodents, and the seed is eaten by birds and small to midsize mammals. It is a very popular ornamental tree due to its brilliant fall foliage, good form, and pest resistance, but it requires a cool site. Cultivars have been developed to increase stress tolerance.

Acer is Latin for "maple tree" and refers to the hardness of the wood; *saccharum* means "sugary" and refers to the sap.

SIMILAR SPECIES Black maple (*Acer nigrum* Michx.) is similar to sugar maple, but black maple has been reported in only one county in the northeastern portion of Alabama. It is identified by leaves that are mostly three-lobed, wilted or droopy looking, and densely hairy on the underside.

Sapodilla Family (*Sapotaceae*)

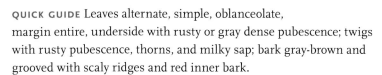

Gum Bumelia

Sideroxylon lanuginosum Michx.

COMMON NAMES gum bumelia, gum bully, woolly buckthorn

QUICK GUIDE Leaves alternate, simple, oblanceolate, margin entire, underside with rusty or gray dense pubescence; twigs with rusty pubescence, thorns, and milky sap; bark gray-brown and grooved with scaly ridges and red inner bark.

DESCRIPTION Leaves are alternate, simple, tardily deciduous, oblance-olate or elliptical or obovate, and 2.5–9 cm (1.0–3.5 in) long; apex is mostly rounded; base is cuneate to rounded; margin is entire; petiole is densely pubescent; underside has velvety rusty or gray pubescence. Twigs are gray-brown with rusty or gray pubescence and thorns and exude milky sap when cut; leaf scar is small and semicircular, with three bundle scars. A true terminal bud is lacking; lateral bud is embedded; scales are overlapping with rusty pubescence. Flowers are perfect and white and bloom in small clusters on maroon pubescent stalks from leaf axils in midsummer. Fruit is a berry, ovoid, about 8 mm (0.3 in) long, pubescent, and black, and matures in the fall. Bark is gray-brown and shallowly grooved on small trees; large trees have scaly ridges and red inner bark. The growth form is a shrub or small tree up to 12 m (40 ft) in height.

Clockwise from left:

Gum bumelia leaves and twig with rusty hair.

Gum bumelia flowers.

Gum bumelia immature fruit.

From left to right:

Bark of a small gum bumelia.

Bark of a large gum bumelia.

HABITAT Stream edges and dry sandy or rocky uplands.

NOTES The wood of gum bumelia is reported to have been used for cabinets. The flowers are popular with bees and the fruit is eaten by birds.

Sideroxylon is Greek for "iron wood"; *lanuginosum* means "with soft hair" and refers to the leaves and twigs. The common name refers to the sticky sap or gum that oozes from cuts in the bark.

Buckthorn Bumelia

Sideroxylon lycioides L.

COMMON NAMES buckthorn bumelia, buckthorn bully

QUICK GUIDE Leaves alternate, simple, and oblanceolate, with an entire margin and mostly glabrous underside when mature; twigs glabrous with thorns and milk sap; bark gray-brown and scaly with red inner bark.

DESCRIPTION Leaves are alternate, simple, tardily deciduous, shiny, elliptical to oblanceolate, and 5–15 cm (2.0–5.9 in) long; apex is acute; base is cuneate; margin is entire; underside is mostly glabrous when mature. Twigs are brown-gray and mostly glabrous, with thorns, and exude milky sap when cut; leaf scar is small and semicircular, with three bundle scars. A true terminal bud is lacking; lateral bud is embedded in spur shoots; scales are overlapping and yellow-green. Flowers are perfect and white and bloom in small clusters from leaf axils in midsummer. Fruit is a berry, ovoid or ellipsoid, 1.0–1.5 cm (0.4–0.6 in) long, black, and glabrous, and matures in the fall. Bark of small trees is gray-brown and ridged; bark of large trees is scaly with red inner bark. The growth form is a shrub or small tree up to 15 m (50 ft) in height.

HABITAT A variety of sites, including stream banks and upland forests.

Clockwise from left:

Buckthorn bumelia leaves.

Buckthorn bumelia flowers.

Buckthorn bumelia fruit.

Buckthorn
bumelia bark.

NOTES The wood of buckthorn bumelia is reported to have been used for cabinets. The flowers are popular with bees, and the fruit is eaten by birds.

Sideroxylon is Greek for "iron wood"; *lycioides* means similar to *Lycium*, a genus with spiny shrubs.

Star-Vine Family (*Schisandraceae*)

Anise-Tree

Illicium floridanum J. Ellis

COMMON NAMES anise-tree, Florida anise, stink-bush

QUICK GUIDE Leaves alternate, simple, evergreen, elliptical, margin entire, lateral veins obscure, with an anise odor when cut; flowers with bright red, fringelike petals and anise smell; fruit a star-shaped cluster of follicles persisting over winter.

DESCRIPTION Leaves are alternate, simple, evergreen, elliptical to lanceolate, 7–15 cm (2.8–5.9 in) long, and leathery, with an anise odor when bruised or cut; apex is acute; base is cuneate; margin is entire; lateral veins are obscure; petiole can be red; underside is glandular. Twigs are red to green or gray-brown and glabrous; leaf scar has one bundle scar. The buds are up to 1.5 cm (0.6 in) long and acute, with green or red scales. Flowers are perfect and star shaped, with up to 33 red fringelike petals that emit an aniselike fragrance and bloom in the spring and early summer. Fruit is a follicle in a star-shaped whorl that becomes dry when the fruit matures in the fall. Bark is brown-gray, smooth, or lightly ridged. The growth form is a multistem shrub or small tree up to 9 m (30 ft) in height and can root sprout.

Clockwise from left:
Anise-tree leaves.
Anise-tree flowers.
Anise-tree fruit.

Anise-tree bark.

HABITAT Moist soils, including banks of streams and rivers, creek and river bottoms, moist slopes and ravines, swamp and pond edges, seeps, and hardwood floodplain forests.

NOTES Anise-tree is planted in hedges and as a small ornamental tree. The leaves are reported to be toxic to livestock.

Illicium means "alluring fragrance"; *floridanum* refers to the geographic range.

Figwort Family (*Scrophulariaceae*)

Royal Paulownia

Paulownia tomentosa (Thunb.) Steud.

COMMON NAMES royal paulownia, princess-tree, empress-tree

QUICK GUIDE Leaves opposite, simple, large, heart shaped, tomentose underside; flowers in large purple terminal clusters before the leaves; fruit a nutlike capsule persistent over winter; bark gray-brown and smooth.

DESCRIPTION Leaves are opposite or occasionally whorled, simple, deciduous, heart shaped, and 13–40 cm (5.1–15.7 in) long; apex is acuminate; base is cordate; margin is entire; petiole is long and stout; underside is tomentose; autumn color is yellow. Seedlings often show lobes or coarse teeth on leaves. Twigs are stout, yellow-brown to green-brown, and pubescent, with corky white lenticels; leaf scar is circular, notched at the top, and raised, with bundle scars forming a circle; pith is chambered or hollow. A true terminal bud is lacking; lateral bud is blunt, embedded, and often superposed; flower bud is large, drooping, velvety brown, and conspicuous in late summer through winter. Flowers are perfect, about 5 cm (2 in) long, tubular, pale purple, and fragrant, and bloom in erect terminal clusters before the leaves. Fruit is a capsule, ovoid, nutlike, woody, two-valved, and about 3 cm (1.2 in) long, and releases small winged seeds in the summer; clusters of open capsules persist over the winter. Bark is gray-brown and mostly smooth or lightly ridged, with lenticels. The growth form is up to 15 m (50 ft) in height and very fast growing.

From left to right:

Royal paulownia leaf.

Royal paulownia flowers.

Royal paulownia fruit.

Royal paulownia
bark.

HABITAT Open and disturbed areas.

NOTES Royal paulownia is a fast-growing tree originally from Asia and naturalized throughout the eastern United States. It is planted as an ornamental tree but suffers from poor form and messy leaves and fruit in the fall, and it produces lots of seeds. This species can be weedy or invasive and is often seen on roadsides or in other open areas. The wood is light and strong and is commercially important in Asia where it is used for cabinets, furniture, and tea chests.

Paulownia is for Princess Paulowna of Russia; *tomentosa* refers to the hair on the leaf underside.

Quassia Family (*Simaroubaceae*)

Tree-of-Heaven

Ailanthus altissima (Mill.) Swingle

COMMON NAMES tree-of-heaven, ailanthus

QUICK GUIDE Leaves alternate, pinnately compound, with up to 41 leaflets; leaflets lanceolate, margins with coarse teeth at the base tipped with black glands on the underside; bark gray-brown and mostly smooth; branches stout and curving upward; fruit a twisted samara in clusters.

DESCRIPTION Leaves are alternate, pinnately compound, deciduous, and up to 1 m (3 ft) long, with 11–41 leaflets. Leaflets are lanceolate and 8–15 cm (3.1–5.9 in) long; apex is acuminate; base is cordate or inequilateral; margin has coarse teeth at the base tipped with black glands on the underside; underside is lightly pubescent; autumn color is yellow or red. Crushed leaves emit an unpleasant odor. Twigs are very stout, tan, and downy, with white lenticels; leaf scar is large, broadly U-shaped, raised, and pale, with many bundle scars; twigs emit a strong odor when cut. A true terminal bud is lacking; lateral bud is round, embedded, and pubescent. Flowers are dioecious; pistillate and staminate flowers are small and yellow-white; flowers bloom in terminal clusters in late spring after the leaves; staminate flowers have an unpleasant odor. Fruit is a samara, yellow to brown, twisted, and about 3 cm (1.2 in) long, with the seed in the center; fruit matures

From left to right:

Tree-of-heaven leaf, note the margin teeth at the base of leaflets.

Tree-of-heaven flowers.

Tree-of-heaven fruit.

Tree-of-heaven bark.

in the summer or fall and clusters persist over the winter. Bark is gray-brown and smooth or lightly ridged with vertical striations. The growth form is up to 15 m (50 ft) in height.

HABITAT Open and disturbed areas.

NOTES Tree-of-heaven is originally from China and has naturalized throughout the eastern United States. It is a fast-growing, aggressive colonizer of open sites and is considered an invasive species in many states. It was first planted as an ornamental in the early 1800s because of its tolerance of poor site conditions, but it can easily take over an area due to excessive seed production and root sprouting. The fruit is eaten by birds.

Ailanthus means "reaching to heaven"; *altissima* means "high or tall."

Bladdernut Family (*Staphyleaceae*)

American Bladdernut

Staphylea trifolia L.

COMMON NAMES American bladdernut, bladdernut

QUICK GUIDE Leaves opposite, trifoliately compound; twigs red-green and mottled, with a pair of buds at the tip; flowers white and bell shaped; fruit a bladderlike capsule; bark gray-brown, smooth, streaked.

DESCRIPTION Leaves are opposite, trifoliately compound (ternate), deciduous, and up to 23 cm (9.1 in) long, with three or rarely five leaflets. Leaflets are ovate and 4–10 cm (1.6–3.9 in) long; apex is acuminate; base is rounded; margin is finely serrate; underside is pubescent; autumn color is yellow. Twigs are red-green, mottled or striped, and glaucous, with lenticels; leaf scar is triangular with three bundle scars and a stipule scar on either side of the leaf scar. A true terminal bud is lacking; lateral buds are usually in pairs at the terminal and ovoid; scales are overlapping, green-brown, and glabrous. Flowers are perfect, white, and bell shaped, and bloom in drooping clusters in late spring after the leaves. Fruit is a capsule, 4–6 cm (1.6–2.4 in) long, balloonlike or bladderlike, thin walled, inflated, and papery, and matures in early fall. Bark is gray-brown to black and smooth or lightly ridged with vertical stripes. The growth form is up to 9 m (30 ft) in height and can form thickets.

HABITAT Moist, fertile soils, including floodplains, stream and river banks, north facing slopes, and rich bluffs.

From left to right:

American bladdernut leaf.

American bladdernut flowers.

American bladdernut fruit.

Clockwise from left:

American bladder-nut bark.

Common hop-tree leaf for comparison.

Common hop-tree flowers for comparison.

NOTES American bladdernut is often found with common hoptree (*Ptelea trifoliata* L.), which also has trifoliately compound leaves, but the leaves of common hoptree are alternate and the fruit is a wafer-like samara. American bladdernut is sometimes planted as an ornamental tree.

Staphylea comes from the Greek for "clusters of grapes," referring to the flowers; *trifolia* means "three leaved."

Storax Family (*Styracaceae*)

Carolina Silverbell

Halesia tetraptera J. Ellis

COMMON NAMES Carolina silverbell, snowdrop tree

QUICK GUIDE Leaves alternate, simple, elliptical, margin with small sharp teeth; flowers white and bell shaped; fruit a persistent four-winged drupe; bark on small stems white striped, becoming red-brown to black and scaly on large trees.

DESCRIPTION Leaves are alternate, simple, deciduous, ovate to obovate or elliptical, and 5–15 cm (2.0–5.9 in) long; apex is acuminate; base is acute to rounded; margin has small sharp teeth or is entire; underside has a white pubescence; autumn color is yellow. Twigs are green-brown, possibly with white streaks, and shreddy when older; leaf scar is shield shaped and raised with one crescent-shaped bundle scar. A true terminal bud is lacking; lateral bud is small, reddish, pubescent, and superposed. Flowers are perfect, white, and bell shaped; petals are fused; flowers bloom in showy clusters before and with the leaves. Fruit is a drupe, about 4 cm (1.6 in) long, four-winged, and papery, and matures in early fall. Bark is red-brown to black and smooth with pale stripes on small trees; on large trees the bark is loose and oblong, with brown-gray to black scales and red-brown inner bark. The growth form is up to 30 m (100 ft) in height and 1 m (3 ft) in diameter. Very large trees often show burls on the trunk.

HABITAT Moist soils, a variety of mesic sites

From left to right:

Carolina silverbell leaf.

Carolina silverbell flowers and [inset] fruit.

Clockwise from left:

Carolina silverbell striped bark.

Carolina silverbell scaly bark.

Coarse margin teeth of two-wing silverbell leaf for comparison.

Petals fused only at the base of two-wing silverbell flowers for comparison.

NOTES Forest associates of Carolina silverbell in the northern portion of the state include red maple, sugar maple, yellow buckeye, black birch, American beech, tulip-poplar, magnolias, eastern white pine, black cherry, northern red oak, and eastern hemlock. Associates in the Southern Coastal Plain include boxelder, red maple, American beech, white oak, swamp chestnut oak, and cherrybark oak. The wood is white to red-brown, soft, light, and close-grained, and is used for pulpwood, veneer, cabinetry, and woodenware. Carolina silverbell is planted as an ornamental, and cultivars with pink flowers are available. The seeds are eaten by birds and small mammals, and the flowers are popular with bees.

Halesia is for the plant physiologist Stephen Hales; *tetraptera* means "four-winged."

SIMILAR SPECIES Two-wing silverbell (*Halesia diptera* J. Ellis) is found primarily on the Southern Coastal Plain on moist soils and is distinguished by leaves with irregular and coarse teeth, flowers with petals fused only at the base, and a two-winged fruit.

American Snowbell

Styrax americanus Lam.

COMMON NAMES American snowbell

QUICK GUIDE Leaves alternate, simple, elliptical, margin wavy and irregularly toothed; buds tan, scurfy, and superposed; flowers with white petals fused only at the base and recurved; fruit an unwinged, pubescent drupe; bark gray-brown and mostly smooth.

DESCRIPTION Leaves are alternate, simple, deciduous, obovate or elliptical, and 3–8 cm (1.2–3.1 in) long; apex tapers to a blunt point; base is acute; margin is wavy and irregularly toothed; petiole is pubescent; underside is glabrous or pubescent. Twigs are red-brown and pubescent or glabrous; leaf scar is shield shaped with one crescent-shaped bundle scar. A true terminal bud is lacking; lateral bud is stalked, about 3 mm (0.1 in) long, superposed, naked, tan, scurfy, and pubescent. Flowers are perfect and white; petals are fused only at the base and are recurved; flowers bloom in showy clusters in the spring after the leaves. Fruit is a drupe, round, 6–13 mm (0.2–0.5 in) wide, and pubescent, and matures in early fall. Bark is gray-brown and mostly smooth. The growth form is a shrub or small tree up to 6 m (20 ft) in height.

HABITAT Moist to wet soils, including damp woods, marshes, swamps, ditches, and stream edges.

From left to right:

American snowbell leaves.

American snowbell flowers.

American snowbell fruit.

American snowbell bark.

NOTES American snowbell is planted as an ornamental tree but requires moist soils. The seeds are eaten by birds and small mammals, and the flowers are visited by bees.

Styrax means "resinous gum tree"; *americanus* refers to the New World.

Bigleaf Snowbell

Styrax grandifolius Aiton

COMMON NAMES bigleaf snowbell

QUICK GUIDE Leaves alternate, simple, round, margin entire or irregularly toothed; flowers white and bell shaped, with petals fused only at the base; buds golden brown, fuzzy; fruit an unwinged pubescent drupe; bark of small trees gray-brown with orange stripes.

DESCRIPTION Leaves are alternate, simple, deciduous, oval to round, and 8–16 cm (3.1–6.3 in) long; apex quickly tapers to a sharp point or is rounded; base is rounded; margin has irregular fine or dentate teeth; underside is pubescent. Twigs are red-brown and pubescent or glabrous; leaf scar is shield shaped with one crescent-shaped bundle scar. A true terminal bud is lacking; lateral bud is stalked, about 6 mm (0.2 in) long, naked, golden brown, fuzzy, and superposed. Flowers are perfect, white, and bell shaped; petals are fused only at the base; flowers bloom in showy clusters in the spring after the leaves. Fruit is a drupe, nearly round or ellipsoid, 7–10 mm (0.3–0.4 in) wide, and pubescent, and matures in early fall. Bark is gray to brown with orange stripes and smooth on small trees; bark on large trees is shallowly grooved with somewhat scaly ridges. The growth form is a shrub or small tree up to 3 m (20 ft) in height.

HABITAT Mesic woods, edges of streams, and swamps.

NOTES The seeds of bigleaf snowbell are eaten by birds and small mammals, and the flowers are visited by bees.

Styrax means "resinous gum tree"; *grandifolius* means "big leaf."

From left to right:

Bigleaf snowbell leaf.

Bigleaf snowbell flowers.

Bigleaf snowbell fruit.

Bark of a bigleaf
snowbell tree.

Sweetleaf Family (*Symplocaceae*)

Sweetleaf

Symplocos tinctoria (L.) L'Her.

COMMON NAMES sweetleaf, horse-sugar, dye bush, yellow wood

QUICK GUIDE Leaves alternate, simple, leathery, sweet tasting, margin mostly entire, petiole and midrib bright yellow; flowers in yellow-white clusters in the spring; bark gray-brown, striped, and smooth or shallowly grooved.

DESCRIPTION Leaves are alternate, simple, evergreen or tardily deciduous, elliptical to oblong, 5–15 cm (2.0–5.9 in) long, and leathery; apex is acute; base is cuneate; margin is mostly entire or with fine teeth; petiole and midrib are bright yellow; underside is pubescent or glabrous. Leaf is sweet tasting, especially near the midrib. Twigs are red-brown and glabrous or pubescent, with a chambered pith; leaf scar is shield shaped with one crescent-shaped bundle scar. Terminal bud is ovoid and about 6 mm (0.2 in) long; scales are green-brown, overlapping, and pubescent; flower bud is plump. Flowers are perfect and short-stalked and bloom in yellow-white clusters in the spring before the new leaves. Fruit is a drupe, oblong, 1.0–1.3 cm (0.4–0.5 in) long, and green or brown, and matures in early fall. Bark is gray-brown to green-brown, striped, and smooth or lightly grooved, with warts. The growth form is up to 9 m (30 ft) in height.

From left to right:

Sweetleaf leaves and flower buds.

Sweetleaf flowers.

Sweetleaf fruit and twig.

Sweetleaf bark.

HABITAT Moist soils of upland forests, swamp borders, floodplains, and stream banks, and occasionally on dry sites.

NOTES Sweetleaf is usually a small tree found growing with a variety of other tree and shrub species. A yellow dye for wool was made from the leaves and bark in Colonial times. The leaves are enjoyed by livestock and moderately browsed by white-tailed deer. *Symplocos* refers to the united stamens; *tinctoria* refers to the dye.

Tea Family (*Theaceae*)

Loblolly-Bay

Gordonia lasianthus (L.) J. Ellis

COMMON NAMES loblolly-bay, gordonia

QUICK GUIDE Leaves alternate, simple, evergreen, elliptical, margin with fine teeth; flowers with five white, silky-haired petals in the summer; bark gray-brown and deeply grooved on large trees.

DESCRIPTION Leaves are alternate, simple, evergreen, elliptical, 8–16 cm (3.1–6.3 in) long, leathery, and shiny; apex is acute; base is cuneate; margin has fine teeth; underside is mostly glabrous. Twigs are red-brown to gray and glabrous or pubescent; leaf scar is shield shaped with bundle scars forming a U shape. The terminal bud is ovoid and pubescent. Flowers are perfect, white, about 7 cm (2.8 in) wide, and fragrant, with five fringed and silky-haired petals, and bloom during the summer. Fruit is a capsule, ovoid, and 1–2 cm (0.4–0.8 in) long, with white downy pubescence; capsule splits along five seams and contains winged seeds; fruit matures in early fall. Bark is red-brown to gray and shallowly grooved on small trees; on large trees bark is gray-brown and deeply grooved with thick ridges. The growth form is up to 24 m (80 ft) in height.

HABITAT Acidic, wet soils of shrub bogs, evergreen bay forests, and pocosins.

From left to right:

Loblolly-bay leaf.

Loblolly-bay flower.

Loblolly-bay fruit.

Loblolly-bay bark.

NOTES Loblolly-bay is found growing with Atlantic white-cedar, swamp cyrilla, sweetbay magnolia, swamp tupelo, redbay, pond pine, and pondcypress. The reddish wood is occasionally used for specialty wood items. The bark was used in tanning leather. It is planted as an ornamental for its shiny evergreen foliage and flowers.

Gordonia is for James Gordon, a British nurseryman; *lasianthus* means "flower with hair."

Elm Family (*Ulmaceae*)

Water-Elm

Planera aquatica (Walter) J. F. Gmel.

COMMON NAMES water-elm, planertree

QUICK GUIDE Leaves alternate, simple, ovate, two-ranked, apex acute, margin serrate; fruit with fleshy projections; bark gray-brown and scaly with red inner bark.

DESCRIPTION Leaves are alternate, simple, deciduous, two-ranked, ovate to lanceolate, and 3–7 cm (1.2–2.8 in) long; apex is acute; base is rounded; margin is serrate. Twigs zigzag and are slender, red-brown, and pubescent, with lenticels; leaf scar is triangular and minute with three bundle scars. A true terminal bud is lacking; lateral bud is ovoid to round and 2 mm (0.1 in) long or smaller; scales are red-brown, overlapping, and pubescent. Flowers are perfect and imperfect, yellow-green or white-green, and lack petals; flowers bloom from leaf axils in the spring with the leaves. Fruit is a drupe, 8–13 mm (0.3–0.5 in) long, and misshapen, with fleshy burlike projections, and matures in the spring. Bark is gray-brown, thin, scaly, and shreddy or shaggy, with red inner bark. The growth form is up to 15 m (50 ft) in height with a low crown and forked, vase-shaped trunk.

Clockwise from left:
Water-elm leaves.
Water-elm flowers.
Water-elm fruit.

Water-elm bark.

HABITAT Swamps, bottomlands, and riverbanks.

NOTES Water-elm is an uncommon tree found with baldcypress, pond-cypress, water tupelo, swamp tupelo, swamp laurel oak, and Carolina ash. The wood is not commercially important, but it is sometimes used for boats. The fruit is eaten by waterfowl.

Planera is for Johann Planer, a German botanist in the eighteenth century; *aquatica* refers to the habitat.

Winged Elm

Ulmus alata Michx.

COMMON NAMES winged elm, cork elm, wahoo, small-leaved elm, hard elm

QUICK GUIDE Leaves alternate, simple, elliptical, apex acute, base slightly inequilateral, margin doubly serrate; twigs often with corky wings; fruit an elliptical, pubescent, notched samara; bark gray-brown with flattened or scaly ridges.

DESCRIPTION Leaves are alternate, simple, deciduous, elliptical or lanceolate, and 3–8 cm (1.2–3.1 in) long; apex is acute; base is slightly inequilateral; margin is doubly serrate; upper side is smooth or sometimes scabrous; autumn color is yellow. Twigs slightly zigzag and are slender, red-brown, and mostly glabrous, with lenticels and often with corky wings; leaf scar is semicircular with three or more bundle scars. A true terminal bud is lacking; lateral bud is acute, about 3 mm (0.1 in) long, and divergent; scales are red-brown, overlapping, and mostly glabrous. Flowers are perfect, in yellow-red to purplish clusters, and bloom in the spring before the leaves. Fruit is a samara, elliptical, 5–10 mm (0.2–0.4 in) long, and deeply notched at the apex; margin has dense white pubescence; fruit matures before or with the leaves. Bark of small trees has gray corky or scaly ridges; bark of large trees is gray-brown to red-brown with flattened, possibly scaly ridges. The growth form is up to 15 m (50 ft) in height and 61 cm (23.6 in) in diameter with drooping branches.

From left to right:

Winged elm leaves and winged twig.

Winged elm flowers.

Winged elm fruit

From left to right:

Bark of large winged elm.

Bark of small winged elm.

HABITAT A wide variety of sites such as stream edges, floodplains, open fields, and sandy uplands.

NOTES Forest associates of winged elm are numerous and site dependent. The wood is brown or red-brown, heavy, hard, and shock resistant, and is used for boxes, crates, furniture, posts, fuel, and hockey sticks. The seed is eaten by birds and small mammals.

Ulmus is Latin for "elm tree"; *alata* means "winged," referring to the twigs.

SIMILAR SPECIES **September elm** (*Ulmus serotina* Sarg.) is an uncommon tree found in northern Alabama that flowers in the summer and fruits in the autumn and has twigs with corky wings. The leaves of September elm are scabrous on the upper surface and have coarse teeth on the margin, and the fruit has long silvery hairs.

American Elm

Ulmus americana L.

COMMON NAMES American elm, white elm, water elm

QUICK GUIDE Leaves alternate, simple, obovate, apex acuminate, base greatly inequilateral, margin doubly serrate; upper surface mostly smooth but can be scabrous; twigs and buds mostly glabrous; fruit a round, notched, pubescent samara; bark gray-brown with scaly ridges and brown-white inner layers.

DESCRIPTION Leaves are alternate, simple, deciduous, obovate, ovate or elliptical, and 8–15 cm (3.1–5.9 in) long; apex is acuminate; base is greatly inequilateral; margin is doubly serrate; upper surface is smooth or slightly scabrous (can be very scabrous on saplings); autumn color is yellow. Twigs slightly zigzag and are slender, red-brown, and mostly glabrous; leaf scar is semicircular with three or more bundle scars. A true terminal bud is lacking; lateral bud is ovoid to acute, about 6 mm (0.2 in) long, and divergent; scales are red-black and mostly glabrous. Flowers are perfect and long-stalked and bloom in red-brown clusters in the spring before the leaves. Fruit is a samara, nearly round, about 1 cm (0.4 in) wide, and deeply notched at the apex; margin has a dense white pubescence; fruit matures in the spring with the developing leaves. Bark is gray to gray-brown with scaly ridges, revealing alternating brown and white layers when cut. The growth form is up to 38 m (125 ft) in height and 1.5 m (5 ft) in diameter, with drooping branches and a forked or vase-shaped trunk.

From left to right:

American elm leaf.

American elm flowers.

American elm fruit.

American elm bark.

HABITAT Moist soils, including upland sites but more common on bottomlands, terraces, and stream and swamp edges.

NOTES Forest associates of American elm include boxelder, red maple, silver maple, sugarberry, hackberry, American beech, southern magnolia, sweetbay magnolia, white and green ash, sweetgum, sycamore, eastern cottonwood, and slippery elm. The wood is whitish-gray to brown, moderately heavy, and moderately hard; is used for boxes, crates, pallets, and furniture; and was used in shipbuilding and for wheel hubs. The bark was used by Native Americans to make rope and canoes. The seeds are eaten by wood ducks, wild turkey, and a variety of songbirds. Before Dutch elm disease, American elm was a favorite urban shade tree.

Ulmus is Latin for "elm tree"; *americana* refers to the New World.

Slippery Elm

Ulmus rubra Muhl.

COMMON NAMES slippery elm, red elm, soft elm

QUICK GUIDE Leaves alternate, simple, obovate, apex acuminate, base inequilateral, margin doubly serrate, upper surface and underside scabrous; buds with maroon pubescence; fruit a round, shallowly notched samara with a glabrous margin; bark gray-brown, with flat ridges and brown inner layers.

DESCRIPTION Leaves are alternate, simple, deciduous, obovate to oval or elliptical, and 10–18 cm (3.9–7.1 in) long; apex is acuminate; base is inequilateral; margin is doubly serrate; upper surface and underside are scabrous; autumn color is yellow. Twigs slightly zigzag and are slender, red-brown to gray, and scabrous, with pubescence; leaf scar is semicircular with three or more bundle scars; inner twig bark is slippery and mucilaginous. A true terminal bud is lacking; lateral bud is ovoid, about 6 mm (0.2 in) long, and divergent; scales are red-black with soft red-maroon pubescence. Flowers are perfect and short-stalked and bloom in red-brown to purplish clusters before the leaves; according to Rogers (1965) flowers turn purple only if flowers are dampened by rain and the pigments can diffuse out. Fruit is a samara, round, 1–2 cm (0.4–0.8 in) wide, and shallowly notched at the apex; margin is glabrous but the seed portion is pubescent; fruit matures in the spring as the leaves develop. Bark is gray-brown or red-brown with corky or flattened interlacing ridges, revealing brown

From left to right:

Slippery elm leaves.

Slippery elm flowers.

Slippery elm fruit and twig.

Slippery elm bark.

inner layers when cut. The growth form is up to 21 m (70 ft) in height and 1 m (3 ft) in diameter with a vase shape.

HABITAT A variety of sites including limestone soils but primarily in bottomlands and moist uplands.

NOTES Forest associates of slippery elm include red maple, silver maple, river birch, hickories, sycamore, white oak, chinkapin oak, northern red oak, black oak, and American elm. The wood is gray-white to red-brown, moderately heavy, and moderately hard, and is used for boxes, crates, pallets, and furniture. The inner bark is used as a home remedy for a variety of ailments, including coughs and sore throats. The seed is eaten by birds and small mammals. Slippery elm is susceptible to Dutch elm disease.

Ulmus is Latin for "elm tree"; *rubra* means "red," perhaps referring to the red-maroon hair on the bud or to the wood.

Appendix

Winter Twigs

Rusty Blackhaw
(*Viburnum rufidulum*)

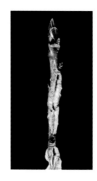

Sweetgum
(*Liquidambar
styraciflua*)

Smoketree
(*Cotinus obovatus*)

Winged sumac
(*Rhus
copallinum*)

Smooth sumac
(*Rhus glabra*)

Poison-sumac
(*Toxicodendron
vernix*)

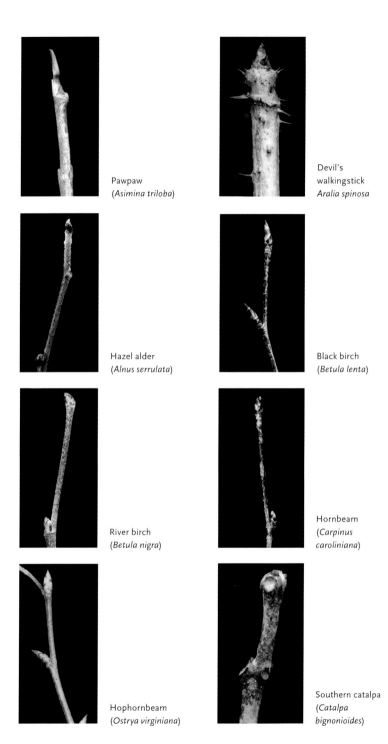

Pawpaw
(*Asimina triloba*)

Devil's
walkingstick
Aralia spinosa

Hazel alder
(*Alnus serrulata*)

Black birch
(*Betula lenta*)

River birch
(*Betula nigra*)

Hornbeam
(*Carpinus
caroliniana*)

Hophornbeam
(*Ostrya virginiana*)

Southern catalpa
(*Catalpa
bignonioides*)

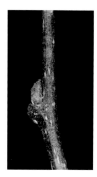

Sugarberry
(*Celtis laevigata*)

Hackberry
(*Celtis occidentalis*)

Georgia hackberry
(*Celtis tenuifolia*)

Alternate-leaf dogwood
(*Cornus alternifolia*)

Flowering dogwood
(*Cornus florida*)

Ogeechee tupelo
(*Nyssa ogeche*)

Swamp tupelo
(*Nyssa biflora*)

Blackgum
(*Nyssa sylvatica*)

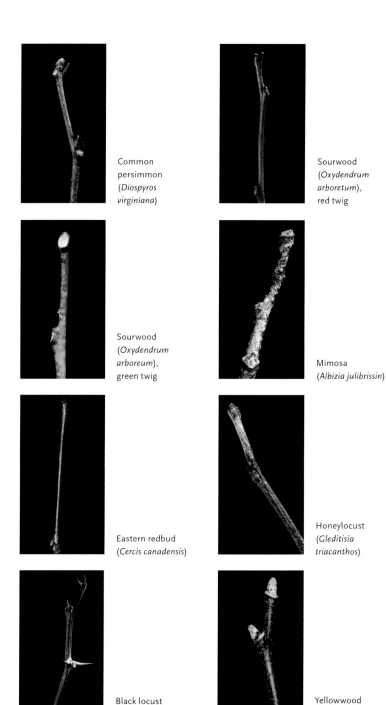

Common
persimmon
(*Diospyros
virginiana*)

Sourwood
(*Oxydendrum
arboretum*),
red twig

Sourwood
(*Oxydendrum
arboreum*),
green twig

Mimosa
(*Albizia julibrissin*)

Eastern redbud
(*Cercis canadensis*)

Honeylocust
(*Gleditisia
triacanthos*)

Black locust
(*Robinia
pseudoacacia*)

Yellowwood
(*Cladrastis
kentukea*)

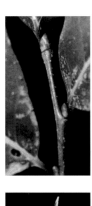

American chestnut
(*Castanea dentata*)

Allegheny
chinkapin
(*Castanea pumila*)

American beech
(*Fagus grandifolia*)

White oak
(*Quercus alba*)

Bluff oak
(*Quercus austrina*)

Scarlet oak
(*Quercus
coccinea*)

Southern red oak
(*Quercus falcata*)

Bluejack oak
(*Quercus incana*)

Turkey oak
(*Quercus laevis*)

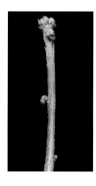

Overcup oak
(*Quercus lyrata*)

Sand post oak
(*Quercus
margarettiae*)

Blackjack oak
(*Quercus
marilandica*)

Swamp chestnut oak
(*Quercus michauxii*)

Chestnut oak
(*Quercus
montana*)

Chinkapin oak
(*Quercus
muehlenbergii*)

Water oak
(*Quercus nigra*)

Cherrybark oak
(*Quercus pagoda*)

Willow oak
(*Quercus
phellos*)

Northern red oak
(*Quercus rubra*)

Shumard oak
(*Quercus
shumardii*)

Post oak
(*Quercus stellata*)

Nuttall oak
(*Quercus texana*)

Black oak
(*Quercus velutina*)

Water hickory
(*Carya aquatica*)

Bitternut hickory
(*Carya cordiformis*)

Pignut hickory
(*Carya glabra*)

Pecan
(*Carya illinoinensis*)

Shellbark hickory
(*Carya laciniosa*)

Nutmeg hickory
(*Carya myristiciformis*)

Shagbark hickory
(*Carya ovata*)

Sand hickory
(*Carya pallida*)

Mockernut
hickory
(*Carya tomentosa*)

Butternut
(*Juglans cinerea*)

Black walnut
(*Juglans nigra*)

Sassafras
(*Sassafras albidum*)

Tulip-poplar
(*Liriodendron
tulipifera*)

Cucumbertree
(*Magnolia acuminata*)

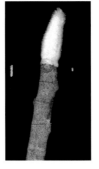

Southern
magnolia
(*Magnolia
grandiflora*)

Bigleaf magnolia
(*Magnolia
macrophylla*)

Umbrella
magnolia
(*Magnolia
tripetala*)

Basswood
(*Tilia americana*)

Chinaberry
(*Melia azedarach*)

Osage-orange
(*Maclura pomifera*)

Red mulberry
(*Morus rubra*)

Fringe-tree
(*Chionanthus virginicus*)

Green ash
(*Fraxinus pennsylvanica*)

Blue ash
(*Fraxinus quadrangulata*)

Sycamore
(*Platanus occidentalis*)

Allegheny service-
berry lateral bud
(*Amelanchier laevis*)

Downy
serviceberry
terminal bud
(*Amelanchier
arborea*)

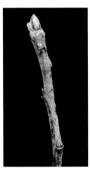

Southern crabapple
(*Malus angustifolia*)

Chickasaw plum
(*Prunus
angustifolia*)

Mexican plum
(*Prunus mexicana*)

Black cherry
(*Prunus serotina*)

Hercules'-club
(*Zanthoxylum
clava-herculis*)

Eastern
cottonwood
(*Populus
deltoides*)

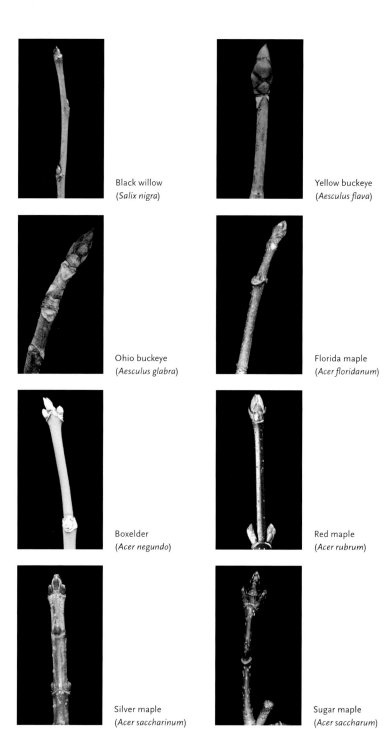

Black willow
(*Salix nigra*)

Yellow buckeye
(*Aesculus flava*)

Ohio buckeye
(*Aesculus glabra*)

Florida maple
(*Acer floridanum*)

Boxelder
(*Acer negundo*)

Red maple
(*Acer rubrum*)

Silver maple
(*Acer saccharinum*)

Sugar maple
(*Acer saccharum*)

Royal paulownia
(*Paulownia
tomentosa*)

Tree-of-heaven
(*Ailanthus
altissima*)

American
bladdernut
(*Staphylea trifolia*)

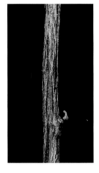

Carolina
silverbell
(*Halesia
tetraptera*)

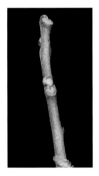

Two-wing silverbell
(*Halesia diptera*)

American
snowbell
(*Styrax
americanus*)

Bigleaf snowbell
(*Styrax grandifolius*)

Water-elm
(*Planera aquatica*)

Winged elm
(*Ulmus alata*)

American elm
(*Ulmus*
americana)

Glossary

ACHENE A dry, unwinged, one-seeded, indehiscent fruit.

ACUMINATE Gradually tapering to an acute angle; attenuated.

ACUTE Forming an angle less than 90 degrees; pointed.

ALTERNATE One leaf at a node.

ANGIOSPERM The flowering plants, with immature seeds enclosed in an ovary.

ANTHER Pollen-bearing structure of the stamen.

APETALOUS Without petals.

APEX The tip of the leaf.

APOPHYSIS The exposed portion of the cone scale when the cone is closed; the "lip" of the cone scale that bears the prickle.

APPRESSED Pressed against.

ARCUATE Archlike, referring to lateral venation that parallels the leaf margin and curves toward the apex.

ARMED Possessing a sharp spine, prickle, or thorn.

ASYMMETRICAL Uneven; unequal in size.

AURICULATE With an earlike appendage, usually referring to the leaf base.

AWL-LIKE Like a pointed tool, tapering to a slender point.

AXIL Angle between the branch and petiole, or midrib and lateral vein.

BERRY A fleshy fruit with one or more seeds.

BIPINNATELY COMPOUND Twice pinnately compound leaf.

BISEXUAL Flower with stamens and pistils; perfect.

BLADE Usually refers to the leaf minus the petiole.

BOG Naturally waterlogged spongy ground, usually composed of a peat and sphagnum moss substrate.

BOTANICAL NAME The Latin binomial name of a plant consisting of the generic name (genus) followed by the specific epithet, which is often descriptive. Developed by Linnaeus in 1753 in *Species Plantarum*.

BOTTOMLAND Low alluvial land, with soil developed from materials recently deposited by streams; subject to frequent flooding.

BRACT A modified leaf associated with a flower or fruit.

BUNDLE SCAR Scars in the leaf scar left by the xylem and phloem.

CAPSULE A dry dehiscent fruit splitting along more than one suture.

CATKINS Apetalous, sessile, unisexual flowers clustered on a single long axis.

CHAMBERED PITH Pith with empty cells.

COMPOUND LEAF A leaf with at least two leaflets.

CORDATE Heart shaped.

CRENATE With rounded teeth on the leaf margin.

CRUCIFORM Crosslike.

CUNEATE The base of a leaf that gradually tapers to an acute angle, wedge shaped.

CYME A flat-topped cluster of flowers blooming from the center outward.

DEHISCENT Splitting along defined sutures or seams.

DELTOID Triangular.

DENTATE Leaf margin with sharp teeth pointing outward.

DIAPHRAGMED PITH A solid pith with cross wall partitions.

DIMORPHIC Having two forms.

DIOECIOUS Having flowers of only one sex on a tree.

DIVERGENT Wide-spread, spread apart.

DOUBLY SERRATE Leaf margin with a smaller tooth within a larger tooth.

DRUPE A fleshy fruit with an inner stony wall enclosing the seed.

ELLIPTICAL Shaped like an ellipse with the widest point in the middle.

EMARGINATE Notched at the leaf apex.

EMBEDDED Buried, enclosed, or fixed.

ENTIRE Leaf margin without teeth or lobes and continuous.

EROSE Eroded, roughly toothed, and jagged.

EXSERTED Visible, extending past the surrounding parts.

FALCATE Sickle shaped.

FASCICLE A bundle of needles.

FILAMENT Threadlike structure of the stamen supporting the anther.

FOLLICLE A dry dehiscent fruit splitting along one suture.

FOUR-RANKED In four vertical rows on an axis.

FURROWED Grooved, plowed.

GLABROUS Smooth, without hair.

GLANDULAR With glands.

GLAUCOUS With a waxy coating.

GLOBOSE Globelike, spherical.

GYMNOSPERM Plants with seeds not borne in an ovary.

HAMMOCK An elevated, well-drained site surrounded by a marsh or bog.

HEAD A round or flat-topped cluster of sessile flowers.

IMBRICATE Overlapping.

IMPERFECT FLOWER Flower of only one sex, with stamens or pistils; unisexual.

INDEHISCENT Not splitting along defined sutures or seams.

INEQUILATERAL Asymmetrical, with unequal sides.

INFLORESCENCE Flower cluster.

KEELED With a prominent longitudinal ridge.

LANCEOLATE Lance shaped, longer than wide, widest below the middle.

LATERAL BUD Axillary bud with a leaf scar, usually smaller than the terminal bud.

LEAF SCAR Scar made by the petiole where it was attached to the branch.

LEGUME A dry podlike fruit splitting along two opposite seams.

LENTICEL A corky bump or line often lens shaped on the surface of a branch or trunk.

LINEAR Long and narrow with parallel sides.

LOBE Rounded part of an organ.

LONG-POINTED With a long point.

LUSTROUS Shiny.

MARGIN The edge.

MESIC Intermediate between wet and dry, moist.

MIDRIB The central vein on the leaf. Usually referring to characteristics of the vein on the underside of the leaf.

MONOECIOUS Having flowers of both sexes on the same tree.

MUCILAGINOUS Slimy.

MUCRONATE A sharp or rigid point.

NAKED BUD Bud lacking visible scales.

NODE Position on the twig where the petiole or branch arises.

NUT A one-seeded fruit with a hard shell.

NUTLET A small nut.

OBLANCEOLATE Reverse lanceolate, widest above the middle.

OBLONG Longer than wide with parallel sides.

OBOVATE Reverse ovate, widest at the apex.

OBTUSE Forming an angle greater than 90 degrees, blunt.

OPPOSITE Two leaves per node on opposite sides of the twig.

ORBICULAR Nearly circular.

OVAL Broadly elliptical.

OVARY The portion of the pistil that encloses the ovules or seeds.

OVATE Egg-shaped outline, widest at the base.

OVOID Egg shaped, three dimensional.

OVULE Immature seed.

PALMATE Arising from a common point, like fingers on a hand.

PANICLE A branched flower structure.

PELTATE Shield shaped, attached from the center.

PENDENT Hanging down.

PERFECT FLOWER A flower with both sexes, with stamens and pistils; bisexual.

PETAL Part of the flower surrounding the reproductive organs, usually white or brightly colored.

PETIOLE The leaf stalk.

PINNATELY COMPOUND A leaf with leaflets arranged along a central axis, like a feather.

PISTIL Female part of a flower made up of an ovary, style, and stigma.

PISTILLATE Female or seed flower.

PITH The spongy center of a twig or root.

POCOSIN An evergreen shrub-tree bog.

POLYGAMOUS Unisexual or bisexual flowers on the same tree.

POME A fleshy fruit with a papery interior wall enclosing the seed.

PRICKLE A sharp or corky outgrowth from the bark, or a sharp structure on a cone scale.

PUBESCENT With soft hairs.

RACEME An unbranched inflorescence, the flowers on a single axis.

RACHIS The main axis of a compound leaf.

RECURVED Curved backward.

REFLEXED Bent downward or backward.

REPAND Irregularly or slightly wavy.

REVOLUTE With the leaf margin rolled under.

RHOMBIC Diamond shaped.

RUGOSE Wrinkled.

SAMARA A dry, winged, indehiscent fruit.

SCABROUS Rough, like sandpaper, to the touch.

SCURFY With small scales.

SEPAL The outer whorl of a flower, usually green but can appear petal-like.

SEROTINOUS Late in flowering or in cone opening.

SERRATE Sharp sawlike teeth on the leaf margin pointing toward the apex.

SESSILE Lacking a petiole or stalk, attached directly.

SHADE LEAF Leaf that formed or developed in shade. Usually refers to leaves located in the lower canopy of trees or leaves of a seedling or sapling grown in the shaded understory.

SHADE TOLERANCE Refers to a plant's vigor in the shaded forest understory and ability to grow from a seedling to an adult under dense cover; may vary with site quality, tree age, and climate.

SIMPLE A leaf with one blade, unbranched, undivided.

SINUATE Wavy.

SINUS Depression or gap between lobes.

SPATULATE Spatula-like with the widest point at the apex, gradually narrowing toward the base.

SPINE A modified leaf or stipule, usually sharp.

STALKED With a supporting structure, like a neck.

STAMEN The male reproductive flower consisting of an anther and filament.

STAMINATE The male or pollen flower bearing the stamens.

STELLATE Star shaped.

STIGMA The top of the style that receives the pollen.

STIPULAR SCAR Scar left on the twig by the stipules.

STIPULE A leafy appendage at the petiole, usually smaller than the leaves and appearing before the leaves.

STRIATIONS Parallel streaks, stripes, or fine grooves.

STYLE Tube connecting the ovary and the stigma.

SUBOPPOSITE Nearly opposite.

SUN LEAF Leaf that formed or developed in full sun. Usually the leaves located in the upper canopy of trees or leaves of an unshaded seedling or sapling.

SUPERPOSED BUD A bud located just above the axillary bud.

SWAMP A seasonally flooded bottomland that is better drained than a bog.

TERMINAL BUD Bud at the tip or end of the shoot, usually larger than the lateral bud, lacking a leaf scar.

THORN A sharp, modified branch.

TOMENTOSE With matted soft or wooly hairs.

TRIFOLIATELY COMPOUND With three leaflets.

TRIPINNATELY COMPOUND A leaf that is pinnately compound three times.

TRUNCATE Squarelike.

TWO-RANKED In two vertical rows on an axis.

UMBEL A flat-topped flower, with flowers arising from the same point and all flower stalks of equal length.

VALVATE Not overlapping, edge to edge.

VEIN AXIL Angle formed between veins, usually referring to axils on the underside of the leaf.

WHORLED With three or more leaves per node.

WINGED With wings or extensions from the seed, flower, rachis, or twig.

WOOLY With long tangled or matted hairs.

Bibliography

Alabama Forestry Commission. 2010. Forests at the crossroads. Alabama's forest assessment and resource strategy. http://www.forestry.alabama.gov/Alabama-ForestActionPlan.aspx?bv=5&s=1.

Alabama Forestry Commission. 2013. Forest resource report 2013. http://www.forestry.alabama.gov/PDFs/alabamaForestResourceReport.pdf.

Angiosperm Phylogeny Group III. 2009. An update of the Angiosperm Phylogeny Group classification of the orders and families of flowering plants: APG III. Botanical Journal of the Linnean Society 161: 105–121.

Bell, C. R., and A. H. Lindsey. 1990. Fall color and woodland harvests. Chapel Hill, N.C.: Laurel Hill Press.

Bishop, G. N. 2000. Native trees of Georgia. Athens: Georgia Forestry Commission.

Brown, C. L., and L. K. Kirkman. 1990. Trees of Georgia and adjacent states. Portland, Ore.: Timber Press.

Burns, R. M., and B. H. Honkala, eds. 1990. Silvics of North America, Volume 1, Conifers. Agriculture Handbook 654. Washington, D.C.: USDA Forest Service.

Burns, R. M., and B. H. Honkala, eds. 1990. Silvics of North America, Volume 2, Hardwoods. Agriculture Handbook 654. Washington, D.C.: USDA Forest Service.

Core, E. L., and N. P. Ammons. 1977. Woody plants in winter. Pacific Grove, Calif.: Boxwood Press.

Davis, D. E., N. D. Davis, and L. J. Samuelson. 1999. Guide and key to Alabama trees. Dubuque, Iowa: Kendall/Hunt Publishing.

Dean, B. E. 1988. Trees and shrubs of the southeast. Birmingham, Ala.: Audubon Society Press.

Dickson, J. G., ed. 2001. Wildlife of southern forests. Blaine, Wash.: Hancock House Publishers.

Dirr, M. A. 1997. Dirr's hardy trees and shrubs. Portland, Ore: Timber Press.

Duncan, W. H., and M. B. Duncan. 1988. Trees of the southeastern United States. Athens: University of Georgia Press.

Farrar, J. L. 1995. Trees of the northern United States and Canada. Ames: Iowa State University Press.

Fralish, J. S., and S. B. Franklin. 2002. Taxonomy and ecology of woody plants in North American forests. New York: John Wiley & Sons, Inc.

Gill, J. L. 2014. Ecological impacts of the late Quaternary megaherbivore extinctions. New Phytologist 201: 1163–1169.

Gill, J. D., and W. M. Healy, eds. 1974. Shrubs and vines for northeastern wildlife.

USDA Forest Service General Technical Report NE-9, Upper Darby, Pa.: Northeastern Forest Experiment Station.

Gledhill, D. 2002. The names of plants. Cambridge, UK: Cambridge University Press.

Godfrey, R. K. 1988. Trees, shrubs, and woody vines of northern Florida and adjacent Georgia and Alabama. Athens: University of Georgia Press.

Godfrey, R. K., and J. W. Wooten. 1981. Aquatic and wetland plants of the southeastern United States. Athens: University of Georgia Press.

Grimm, W. C. Revised by J. Kartesz. 2002. The illustrated book of trees. Mechanicsburg, Pa.: Stackpole Books.

Halls, L. K., ed. 1977. Southern fruit-producing woody plants used by wildlife. USDA General Technical Report SO-16. Upper Darby, Pa.: Southern Forest Experiment Station.

Halls, L. K., and T. H. Ripley. 1961. Deer browse plants of southern forests. U.S. Forest Service Report. Upper Darby, Pa.: South and Southeast Forest Experiment Station.

Hardin, J. W., D. J. Leopold, and F. M. White. 2001. Harlow & Harrar's textbook of dendrology. New York: McGraw-Hill Company, Inc.

Harlow, W. M., E. S. Harrar, J. W. Hardin, and F. M. White. 1996. Textbook of dendrology. New York: McGraw-Hill, Inc.

Harper, R. M. 1928. Economic botany of Alabama. Geological Survey of Alabama Monograph 9, University, Ala.

Harper, R. M. 1943. Forests of Alabama. Geological Survey of Alabama Monograph 10, University, Ala.

Harrar, E. S., and J. G. Harrar. 1962. Guide to southern trees. New York: Dover Publications.

Harris, J. G., and M. W. Harris. 1999. Plant identification terminology. Spring Lake, Utah: Spring Lake Publishing.

Hunter, C. G. 2004. Trees, shrubs & vines of Arkansas. Little Rock, AR: Ozark Society Foundation.

Hutnick, R. J., and H. W. Yawney. 1961. Silvical characteristics of red maple (*Acer rubrum*). USDA Forest Service Station Paper 142. Upper Darby, Pa.: Northeastern Forest Experiment Station.

Keener, B. R., A. R. Diamond, L. J. Davenport, P. G. Davison, S. L. Ginzbarg, C. J. Hansen, C. S. Major, D. D. Spaulding, J. K. Triplett, and M. Woods. 2018. Alabama Plant Atlas. (S. M. Landry and K. N. Campbell [original application development], Florida Center for Community Design and Research. University of South Florida.). University of West Alabama, Livingston, Ala. www.floraofalabama.org.

Leopold, D. J., W. C. McComb, and R. N. Muller. 1998. Trees of the central hardwood forests of North America. Portland, Ore.: Timber Press.

Linnaeus, C. 1753. Species plantarum. Stockholm, Sweden.

Little, E. L., Jr. 1971. Atlas of United States trees, Vol. 1. Conifers and important hardwoods. Miscellaneous Publication Number 1146. Washington D.C.: USDA Forest Service.

Little, E. L., Jr. 1977. Atlas of United States trees, Vol. 4. Minor eastern hardwoods. Miscellaneous Publication Number 1342. Washington D.C.: USDA Forest Service.

Little, E. L., Jr. 1979. Checklist of United States trees (native and naturalized). Agriculture Handbook No. 541. Washington, D.C.: USDA Forest Service

Martin, A. C., H. S. Zinn, and A. L. Nelson. 1951. American wildlife and plants: A guide to wildlife food habits. New York: McGraw-Hill Book Company, Inc.

Matoon, W. R. 1948. Common forest trees of Florida. Tallahassee: Florida Board of Forestry and Parks.

Miller, H. A., and S. H. Lamb. 1985. Oaks of North America. Happy Camp, Calif.: Naturegraph Publishers, Inc.

Miller, J. H., and K. V. Miller. 1999. Forest plants of the southeast and their wildlife uses. Champagne, Ill.: Southern Weed Society.

Panshin, A. J., and C. de Zeeuw. 1980. Textbook of wood technology. New York: McGraw-Hill Book Company, Inc.

Preston, R. J., and R. R. Braham. 2002. North American trees. Ames: Iowa State Press.

Rogers, J. E. 1920. The tree book: A popular guide to a knowledge of the trees of North America and to their uses and cultivation. New York: Doubleday, Page and Company.

Rogers, W. E. 1965. Tree flowers of forest, park and street. New York: Dover Publications.

Samuelson, L. J., and M. E. Hogan. 2003. Forest trees: A guide to the southeastern and mid-Atlantic regions of the United States. Upper Saddle River, N.J.: Prentice Hall.

Samuelson, L. J., and M. E. Hogan. 2006. Forest trees: A guide to the eastern United States. Upper Saddle River, N.J.: Prentice Hall.

Seiler, J. R., and J. A. Peterson. Dendrology at Virginia Tech Web Site. http://dendro.cnre.vt.edu/dendrology/main.htm.

Settergren, C., and R. E. McDermott. 1995. Trees of Missouri. University Extension Publication SB 767. Columbia, Mo.: University of Missouri Press.

Sharitz, R. R., and J. W. Gibbons. 1982. The ecology of southeastern shrub bogs (pocosins) and Carolina bays: A community profile. Publication No. FWS/OBS-82/04. Washington, D.C.: U.S. Fish and Wildlife Service, Division of Biological Services.

Sternberg, G., and J. Wilson. 2004. Native trees for North American landscapes. Portland, Ore.: Timber Press.

Stupka, A. 1993. Trees, shrubs and woody vines of Great Smoky Mountains National Park. Knoxville: University of Tennessee Press.

USDA, NRCS. 2013. The PLANTS Database. Greensboro, N.C.: National Plant Data Team. http://plants.usda.gov.

Vankat, J. L. 1979. The natural vegetation of North America. New York: John Wiley & Sons, Inc.

Vitale, A. T. 1997. Leaves in myth, magic and medicine. New York: Stewart, Tabori & Chang.

Wagner, W. H. 1979. Modern carpentry. South Holland, Ill.: Goodheart-Willcox Co., Inc.

Weakley, A. S. 2012. Flora of the southern and mid-Atlantic states. Chapel Hill: University of North Carolina Herbarium, North Carolina Botanical Garden.

Whittemore, A. T., and K. C. Nixon. 2005. Proposal to reject the name *Quercus prinus* (Fagaceae). Taxon 54: 213–214.

York, H. H. 1999. 100 Forest Trees of Alabama. Auburn: Alabama Forestry Commission and State Department of Education, Office of Career/Technical Education, Agriscience Technology Education.

Index

Page numbers in italics refer to illustrations.

Acer floridanum. See Florida Maple
Acer leucoderme. See Chalk Maple
Acer negundo. See Boxelder
Acer nigrum. See Black Maple
Acer rubrum var. *drummondi. See* Drummond Maple
Acer rubrum var. *rubrum. See* Red Maple
Acer saccharinum. See Silver Maple
Acer saccharum. See Sugar Maple
achene, 323
acuminate
  defined, 323
  leaf apex, *8*
acute
  defined, 323
  leaf apex, *8*
  leaf base, *8*
Adoxaceae. See Moscatel Family
Aesculus flava. See Yellow Buckeye
Aesculus glabra. See Ohio Buckeye
Aesculus parviflora. See Bottlebrush Buckeye
Aesculus pavia. See Red Buckeye
Aesculus sylvatica. See Painted Buckeye
ailanthus. *See* Tree-of-Heaven
Ailanthus altissima. See Tree-of-Heaven
Alabama Black Cherry or Alabama Chokeberry (*Prunus alabamensis*), 256
Alabama Plant Atlas (Keener et al.), 2
Albizia julibrissin. See Mimosa
Albizzi, Filippo, 120
Allegheny Chinkapin (*Castanea pumila*), 20, 129–130
  bark (courtesy of Nancy Loewenstein), *130*
fruit (courtesy of Alan Cressler), *129*
  leaves, *129*
  winter twig, *313*
Allegheny Serviceberry (*Amelanchier laevis*), 244
  lateral bud, *319*
Alnus serrulata. See Hazel Alder
alternate leaf arrangement, 3–4, 12–17
  Alternate-Leaf Dogwood, 19, *101*
  defined, 323
Alternate-Leaf Dogwood (*Cornus alternifolia*), 19, 101–102
  bark, *102*
  flowers, *101*
  fruit, *101*
  leaves, *101*
  winter twig, *311*
Altingiaceae. See Sweetgum Family
Amelanchier arborea. See Downy Serviceberry
Amelanchier laevis. See Allegheny Serviceberry
American Beech (*Fagus grandifolia*), 20, 131–132
  bark, *132*
  flowers, *131*
  fruit, *131*
  with winter leaves, *132*
  winter twig, *313*
American Bladdernut (*Staphylea trifolia*), 289–290
  bark, *290*
  common hoptree flowers for comparison, *290*
  common hoptree leaf for comparison, *290*
  flowers, *289*
  fruit, *289*
  leaf, *289*

winter twig, *321*

American Chestnut (*Castanea dentata*), 20, 127–128
  bark of sapling, *128*
  Chinese Chestnut leaf with smaller margin teeth for comparison, *128*
  fruit, *127*
  leaves, *127*
  winter twig, *313*

American Elm (*Ulmus americana*), 30, 305–306
  bark, *306*
  flowers, *305*
  fruit, *305*
  leaf, *305*
  winter twig, *322*

American Holly (*Ilex opaca*), 17, 75–76
  bark, *75–76*
  leaves and fruit, 17, *75*
  staminate flowers, *75*

American hornbeam. *See* Hornbeam

American persimmon. *See* Common Persimmon

American planetree. *See* Sycamore

American Plum (*Prunus americana*), *250*

American Smoketree (*Cotinus obovatus*), 17, 61–62
  bark, *62*
  flowers, *61*
  fruit, 61–62
  leaves, 17, *61*

American Snowbell (*Styrax americanus*), 29–30, 293–294
  bark, *294*
  flowers, *293*
  leaves, *293*
  winter twig, *321*

American sycamore. *See* Sycamore

American walnut. *See* Black Walnut

Anacardiaceae. See Cashew Family

angelica-tree. *See* Devil's Walkingstick

Angiosperms
  defined, 323
  guide to, 11–17

identification of using leaves, bark, and leaf scars, 3–5

leaves alternate and compound: twigs armed, 15–16
  leaves bipinnately compound, 16
  leaves pinnately compound, 15–16

leaves alternate and compound: twigs unarmed, 16–17
  leaves bipinnately compound, 16–17
  leaves pinnately compound, 16

leaves alternate, simple, and lobed, 12

leaves alternate, simple, and unlobed: fragrant when crushed, 12

leaves alternate, simple, and unlobed: margins without teeth, 12–13
  leaves deciduous
    leaves heart shaped or triangular, 13
    leaves neither heart shaped nor triangular, 13
  leaves evergreen, 12–13

leaves alternate, simple, and unlobed: margins with teeth, 14–15
  leaves deciduous and heart shaped or triangular, 14
  leaves deciduous with doubly serrate margins, 15
  leaves deciduous with singly serrate or irregularly toothed margins, 14–15
  leaves tardily deciduous or evergreen, 14

leaves alternate, simple, and unlobed: twigs armed, 12

leaves neither heart shaped nor triangular, 13

leaves opposite and compound, 11

leaves opposite or whorled and simple, 11

Anise-Tree (*Illicium floridanum*), 283–284
  bark, *284*
  flowers, *283*
  fruit, *283*
  leaves, *283*

Annonaceae. See Custard-Apple Family

anther, 323

apetalous, 323

apex

  acuminate, Dwarf Pawpaw, *70*

  defined, 323

  nut, Scarlet Oak, *137*

  with and without bristle tip, Beech
    Family, 21–23

apophysis, 323

appressed, 323

Aquifoliaceae. See Holly Family

Araliaceae. See Ginseng Family

Aralia spinosa. See Devil's Walkingstick

arcuate, 323

Arkansas Oak (*Quercus arkansana*),
  156

armed, 12, 15–16, 323

ashleaf maple. *See* Boxelder

Asimina parviflora. See Dwarf Pawpaw

Asimina triloba. See Pawpaw

asymmetrical, 323

Atlantic White-Cedar (*Chamaecyparis
  thyoides*), 9, 31–32

  bark on large tree, *32*

  bark on young, *32*

  leaves, *31*

  seed cones, *31*

auriculate

  defined, 323

  leaf base, *8*

awl-like, 323

axil, 323

Baldcypress (*Taxodium distichum*), 9,
  37–38

  bark, *37–38*

  "cypress knees," *37*

  fluted trunk, *38*

  leaves, *37*

  seed cone, *37*

bark

  of American Chestnut sapling, *128*

  blocky, of Blackgum, 19, *110*

  with brown flecking, of Slash Pine, *42*

  with cracks, of Downy Serviceberry,
    244

  exposed pale bark, of Sycamore, *240*

  flaky, of Cherrybark Oak, *166*

  with flat ridges, of Scarlet Oak, *138*

  with flattened ridges, of Blackgum, *110*

  furrowed (*See* furrowed bark)

  golden, of young Yellow Birch, *84*

  grooved, of White Oak, *134*

  juvenile versus mature, 4

  and "knees," cypress, *36, 37*

  of large Atlantic White-Cedar, *32*

  of large Basswood, *222*

  of large Black Birch, *84*

  of large Black Cherry, *256*

  of large Bluejack Oak, *146*

  of large Devil's Walkingstick, *80*

  of large Eastern Hemlock, *56*

  of large Gallberry, *71–72*

  of large Gum Bumelia, *280*

  of large Laurel Oak, *144*

  of large Loblolly Pine, *52*

  of large Northern Red Oak, *169–170*

  of large Post Oak, *174*

  of large Tulip-Poplar, *210*

  of large Virginia Pine, *54*

  of large Water Oak, *164*

  of large Willow Oak, *167–168*

  of large Winged Elm, *304*

  loose plates of, on Black Willow, *262*

  loose plates of, on White Oak, *134*

  orange-gray scaly, of Sand Pine, *40*

  with orange in grooves, of Chestnut
    Oak, *160*

  with orange in grooves, of Common
    Persimmon, *112*

  with pale flecking, of Slash Pine, *42*

  peeling, of River Birch, *86*

  with resin holes in, of Shortleaf Pine,
    40

  scaly (*See* scaly bark)

  shaggy, of Water Hickory, *182*

  of small Bigleaf Snowbell, *296*

  of small Butternut, *202*

  of small Chalk Maple, *270*

of small Gum Bumelia, *280*
of small Winged Elm, *304*
smooth, of young Eastern White Pine, *50*
smooth or shallowly ridged, of small Spruce Pine, *44*
with sprouts, of Pond Pine, *48*
striped, of Carolina Silverbell, *292*
striped, of Downy Serviceberry, *244*
thick, of large Eastern Hemlock, *56*
with thorns, of small Honeylocust, *124*
use of to identifiy trees, *4*
with white "ski tracks," of Northern Red Oak, *169–170*
with white streaks, of Scarlet Oak, *138*
of young Atlantic White-Cedar, *32*
of young Black Birch, *84*
of young Devilwood, *238*
of young Eastern Hemlock, *56*
of young Laurel Oak, *144*
of young Loblolly Pine, *52*
of young Mockernut Hickory, *200*
of young Nutmeg Hckory, *192*
of young Redbay, *206*
of young Tulip-Poplar, *210*
of young Virginia Pine, *54*
basket oak. *See* Swamp Chestnut Oak
Basswood (*Tilia americana*), 221–222
  bark of large, *222*
  flowers, *221*
  fruit, *221*
  leaves, *221*
  winter twig, *318*
bay pine. *See* Pond Pine
Bean or Pea Family (*Fabaceae*), 13, 15, 16–17, 119–126
  Black Locust (*Robinia pseudoacacia*), 19, 125–126
  Eastern Redbud (*Cercis canadensis*), 19, 121–122
  guide to, 19–20
  Honeylocust (*Gleditsia triacanthos*), 19, 123–124
  Mimosa (*Albizia julibrissin*), 20, 119–120

Beech Family (*Fagaceae*), 12–13, 15, 127–180
  Allegheny Chinkapin (*Castanea pumila*), 20, 129–130
  American Beech (*Fagus grandifolia*), 20, 131–132
  American Chestnut (*Castanea dentata*), 20, 127–128
  Blackjack Oak (*Quercus marilandica*), 20, 155–156
  Black Oak (*Quercus velutina*), 21, 177–178
  Bluejack Oak (*Quercus incana*), 21, 145–146
  Bluff Oak (*Quercus austrina*), 22, 135–136
  Cherrybark Oak (*Quercus pagoda*), 21, 165–166
  Chestnut Oak (*Quercus montana*), 23, 159–160
  Chinkapin Oak (*Quercus muehlenbergii*), 23, 161–162
  Durand Oak (*Quercus durandii*), 23, 139–140
  fruit an acorn; branch terminal with a cluster of buds, 20–23
    lobed red oaks: lobes bristle tipped, 20–21
    lobed white oaks: lobes and apex lacking bristle tips, 22–23
    unlobed red oaks: apex with bristle tip, 21–22
    unlobed white oaks: leaves of most species lacking bristles at apex and margin, some with scalloped margins, 23
  fruit is nut in spiny bur; branch terminal with a single bud, 20
  guide to, 20–23
  Laurel Oak (*Quercus hemisphaerica*), 21–22, 143–144
  Live Oak (*Quercus virginiana*), 23, 179–180
  Northern Red Oak (*Quercus rubra*), 21, 169–170

Nuttall Oak (*Quercus texana*), 21, 175–176
Overcup Oak (*Quercus lyrata*), 22, 151–152
Post Oak (*Quercus stellata*), 22, 173–174
Sand Post Oak (*Quercus margarettiae*), 22, 153–154
Scarlet Oak (*Quercus coccinea*), 21, 137–138
Shumard Oak (*Quercus shumardii*), 21, 171–172
Southern Red Oak (*Quercus falcata*), 20, 141–142
Swamp Chestnut Oak (*Quercus michauxii*), 23, 157–158
Swamp Laurel Oak (*Quercus laurifolia*), 22, 149–150
Turkey Oak (*Quercus laevis*), 20, 147–148
Water Oak (*Quercus nigra*), 22, 163–164
White Oak (*Quercus alba*), 22–23, 133–134
Willow Oak (*Quercus phellos*), 22, 167–168
berry, 323
Betulaceae. See Birch Family
Betula alleghaniensis. See Yellow Birch
Betula lenta. See Black Birch
Betula nigra. See River Birch
big buckeye. *See* Yellow Buckeye
big bud hickory. *See* Mockernut Hickory
Bigleaf Magnolia (*Magnolia macrophylla*), 26, 215–216
  bark, 216
  flower, 215
  fruit, 215
  leaves, 215
  winter twig, 317
bigleaf shagbark hickory. *See* Shellbark Hickory
Bigleaf Snowbell (*Styrax grandifolius*), 30, 295–296
  bark, 296
  flowers, 295
  fruit, 295
  leaf, 295
  winter twig, 321
Bignoniaceae. See Catalpa Family
big shagbark hickory. *See* Shellbark Hickory
bigtree plum. *See* Mexican Plum
bipinnately compound leaf, 4, 16–17, 323
Birch Family (*Betulaceae*), 14, 15, 81–90
  Black Birch (*Betula lenta*), 18, 83–84
  guide to, 18
  Hazel Alder (*Alnus serrulata*), 18, 81–82
  Hophornbeam (*Ostrya virginiana*), 18, 89–90
  Hornbeam (*Carpinus caroliniana*), 18, 87–88
  River Birch (*Betula nigra*), 18, 85–86
bisexual, 323, 326
Bitternut Hickory (*Carya cordiformis*), 25, 183–184
  bark, 183–184
  fruit, 183
  leaf, 183
  leaf form from underside, 183
  winter twig, 316
bitter pecan. *See* Water Hickory
Black Birch (*Betula lenta*), 18, 83–84
  bark of large tree, 84
  bark of young tree, 84
  golden bark of a young Yellow Birch for comparison, 84
  leaves and fruit, 83
  winter twig, 310
Black Cherry (*Prunus serotina*), 28, 255–256
  bark of large tree, 256
  flowers, 255
  fruit, 255
  leaves, 255
  scaly bark of, 256
  winter twig, 319
black cypress. *See* Pondcypress
black drink holly. *See* Yaupon

Blackgum (*Nyssa sylvatica*), 19, 109–110
  bark, with flattened ridges, *110*
  blocky bark, 19, *110*
  fruit, *109–110*
  pistillate flowers, *109–110*
  staminate flowers, *109–110*
  winter twig, *311*
Blackjack Oak (*Quercus marilandica*),
  20, 155–156
  bark, *156*
  fruit, *155*
  leaves, *155*
  winter twig, *314*
Black Locust (*Robinia pseudoacacia*), 19,
  125–126
  bark, *125–126*
  flowers and fruit, *125*
  leaf, *125*
  obovate leaflets of a Yellowwood leaf
    for comparison, *126*
  winter twig, *312*
Black Maple (*Acer nigrum*), 278
black oak. *See* Blackjack Oak; Black
  Oak
Black Oak (*Quercus velutina*), 21,
  177–178
  bark, *177–178*
  fruit, *177*
  shade leaves, *177*
  sun leaf, *177*
  winter twig, *315*
Black Walnut (*Juglans nigra*), 24,
  203–204
  bark, *204*
  fruit, *203*
  leaf, *203*
  nut, *203*
  winter twig, *317*
Black Willow (*Salix nigra*), 28, 261–262
  bark, *262*
  flowers, *261*
  fruit with cottonlike seeds, *261*
  leaves, *261*
  loose plates of bark, *262*
  winter twig, *320*
bladdernut. *See* American Bladdernut

Bladdernut Family (*Staphyleaceae*), 11,
  289–290
    American Bladdernut (*Staphylea trifo-
    lia*), 289–290
blade, 323
Blue Ash (*Fraxinus quadrangulata*), 236
  winter twig, *318*
blue beech. *See* Hornbeam
blue haw. *See* Rusty Blackhaw
Bluejack Oak (*Quercus incana*), 21,
  145–146
  bark of large tree, *146*
  fruit, *145*
  leaves, *145*
  revolute margins of Myrtle Oak leaves
    for comparison, *146*
  winter twig, *313*
Bluff Oak (*Quercus austrina*), 22,
  135–136
  bark, *136*
  fruit, *135*
  leaves with little lobing, *135*
  leaves with lobes, *135*
  winter twig, *313*
bodark. *See* Osage-Orange
bog, 323
bois d'arc. *See* Osage-Orange
botanical name, 323
Bottlebrush Buckeye (*Aesculus parviflora*), 264
bottomland, 324
bottomland red oak. *See* Cherrybark
  Oak
bow-wood. *See* Osage-Orange
Boxelder (*Acer negundo*), 1, 29, 271–272
  bark, *272*
  fruit, *271*
  leaf, *271*
  pistillate flowers, *271*
  staminate flowers, *271*
  winter twig, *320*
bract, 324
brookside alder. *See* Hazel Alder
Broussonetia papyrifera. *See*
  Paper-Mulberry
buckthorn bully. *See* Buckthorn

Bumelia
Buckthorn Bumelia (*Sideroxylon lycioides*), 29, 281–282
  bark, *282*
  flowers, *281*
  fruit, *281*
  leaves, *281*
Buckthorn Family (*Rhamnaceae*), 14, 241–242
  Carolina Buckthorn (*Frangula caroliniana*), 241–242
buds
  characteristics, 5
  hair on, 5
  lateral, 5, *319*, 325
  naked, 5, 326
  pseudoterminal, 5
  superposed, 328
  terminal, 5, 45–46, *319*, 328
bud scales, 5
bull bay. *See* Southern Magnolia
bundle scar, 5, 324
Bur Oak (*Quercus macrocarpa*), lobing and sinuses of leaves and acorn and cap, *152*
butternut canker, 202
Butternut (*Juglans cinerea*), 23–24, 201–202
  bark of small, *202*
  fruit, *201*
  leaf, *201*
  nut, *201*
  winter twig, *317*
buttonball-tree. *See* Sycamore
buttonwood. *See* Sycamore

Calycanthaceae. *See* Strawberry-Shrub Family
Calycanthus floridus. *See* Sweetshrub
Canadian hemlock. *See* Eastern Hemlock
candleberry. *See* Southern Bayberry
Cannabaceae. *See* Cannabis and Hop Family
Cannabis and Hop Family (*Cannabaceae*), 14–15, 95–100

Georgia Hackberry (*Celtis tenuifolia*), 18, 99–100
  guide to, 18
  Hackberry (*Celtis occidentalis*), 18, 97–98
  Sugarberry (*Celtis laevigata*), 18, 95–96
capsule, 324
Carolina allspice. *See* Sweetshrub
Carolina Ash (*Fraxinus caroliniana*), 236
Carolina basswood. *See* Basswood
Carolina Buckthorn (*Frangula caroliniana*), 241–242
  bark, *242*
  fruit and twig, *241*
  leaves and fruit, *241*
Carolina Hemlock (*Tsuga carolinina*), 55, 56
Carolina Holly (*Ilex ambigua*), 74
Carolina laurel cherry. *See* Cherry Laurel
Carolina Silverbell (*Halesia tetraptera*), 29, 291–292
  coarse margin teeth of two-wing silverbell leaf for comparison, *292*
  flowers and [inset] fruit, *291*
  leaf, *291*
  petals fused only at the base of two-wing silverbell flowers for comparison, *292*
  scaly bark, *292*
  striped bark, *292*
  winter twig, *321*
Carpinus caroliniana. *See* Hornbeam
Carya aquatica. *See* Water Hickory
Carya carolinae-septentrionalis. *See* Southern Shagbark Hickory
Carya cordiformis. *See* Bitternut Hickory
Carya glabra. *See* Pignut Hickory
Carya Illinoinensis. *See* Pecan
Carya laciniosa. *See* Shellbark Hickory
Carya myristiciformis. *See* Nutmeg Hickory
Carya ovalis. *See* Red Hickory
Carya ovata. *See* Shagbark Hickory
Carya pallida. *See* Sand Hickory

Carya tomentosa. See Mockernut Hickory

Cashew Family (*Anacardiaceae*), 13, 16, 61–68
    American Smoketree (*Cotinus obovatus*), 17, 61–62
    guide to, 17
    Poison-Sumac (*Toxicodendron vernix*), 17, 67–68
    Smooth Sumac (*Rhus glabra*), 17, 65–66
    Winged Sumac (*Rhus copallinum*), 17, 63–64

Castanea dentata. See American Chestnut

Castanea mollissima. See Chinese Chestnut

Castanea pumila. See Allegheny Chinkapin

Catalpa bignonioides. See Southern Catalpa

Catalpa Family (*Bignoniaceae*), 11, 91–92
    Southern Catalpa (*Catalpa bignonioides*), 91–92

Catalpa speciosa. See Northern Catalpa

catalpa sphinx larva (catalpa worms), 92

catawba. *See* Southern Catalpa

catkins, 324

cedar pine. *See* Spruce Pine

Celtis laevigata. See Sugarberry

Celtis occidentalis. See Hackberry

Celtis tenuifolia. See Georgia Hackberry

Cercis canadensis. See Eastern Redbud

Chalk Maple (*Acer leucoderme*), 29, 269–270
    bark of small tree, 270
    fruit, *269*
    leaf, *269*
    leaves, *269*

Chamaecyparis thyoides. See Atlantic White-Cedar

chambered pith, 324

Cherrybark Oak (*Quercus pagoda*), 21, 165–166

flaky bark of, 165–166
    fruit, *165*
    furrowed bark of, *166*
    shade leaves of, *165*
    sun leaves of, *165*
    winter twig, *315*

cherry birch. *See* Black Birch

Cherry Laurel (*Prunus caroliniana*), 251–252
    bark, *252*
    flowers, *251*
    fruit, *251*
    leaves (note margin teeth), *251*

chestnut blight, 128

Chestnut Oak (Quercus montana), 23, 159–160
    bark, *160*
    bark with orange in grooves, *160*
    fruit, *159*
    leaves, *159*
    winter twig, *314*

Chickasaw Plum (*Prunus angustifolia*), 27, 249–250
    bark, *250*
    flowers, *249*
    fruit, *249*
    leaves, *249*
    winter twig, *319*
    young tree with thorns, *250*

Chinaberry (*Melia azedarach*), 223–224
    bark, *224*
    flowers, *223*
    fruit, *223*
    fruit in winter, *224*
    leaf, *223*
    winter twig, *318*

Chinese Chestnut (*Castanea mollissima*), leaf with margin teeth, *128*

Chinese Tallowtree (*Triadica sebifera*), 117–118
    bark, *118*
    fruit, *117–118*
    leaves and flowers, *117*

Chinese umbrella tree. *See* Chinaberry

Chinkapin Oak (*Quercus*

muehlenbergii), 23, 161–162
  bark, *162*
  fruit, *161–162*
  leaf margin with callous-tipped teeth,
    161
  leaves, *161*
  obovate leaf, *161*
  winter twig, *314*
chinquapin. *See* Allegheny Chinkapin
Chionanthus virginicus. See Fringe-Tree
chittamwood. *See* American Smoketree
cigar tree. *See* Southern Catalpa
cinnamon wood. *See* Sassafras
Cladrastis kentukea. See Yellowwood
common alder. *See* Hazel Alder
common dogwood. *See* Flowering
  Dogwood
common hackberry. *See* Hackberry
Common Hoptree (*Ptelea trifoliata*)
  flowers, *290*
  leaf, *290*
common pawpaw. *See* Pawpaw
Common Persimmon (*Diospyros virgin-*
  iana), 111–112
  bark, *112*
  bark with orange in the grooves, *112*
  fruit, *111–112*
  leaves and flowers, *111*
  winter twig, *312*
common tree names, 1
compound leaf, 11, 15–17, 324
cone
  flat or peltate scales, 3
  identification of trees based on, 3
  See also seed cones
cordate
  defined, 324
  leaf base, *8*
cork elm. *See* Winged Elm
Cornaceae. See Dogwood Family
Cornus alternifolia. See Alternate-Leaf
  Dogwood
Cornus amomum. See Silky Dogwood
Cornus asperifolia. See Eastern Roughleaf Dogwood

Cornus florida. See Flowering Dogwood
Cornus foemina. See Swamp Dogwood
Cotinus obovatus. See American
  Smoketree; Smoketree
cotton-gum. *See* Water Tupelo
cow oak. *See* Swamp Chestnut Oak
Crataegus spp. *See* Hawthorns
crenate leaf margin, *7*
cruciform, *173*, 324
cucumber magnolia. *See*
  Cucumbertree
Cucumbertree (*Magnolia acuminata*),
  26, 211–212
  bark, *212*
  flower, *211*
  fruit with stinkbug on it, *211*
  leaves, *211*
  winter twig, *317*
cuneate leaf base, *8*, 324
Cupressaceae. See Cypress Family
Custard-Apple Family (*Annonaceae*),
  13, 69–70
  Pawpaw (*Asimina triloba*), 69–70
cyme, 324
Cypress Family (*Cupressaceae*), 31–38
  Atlantic White-Cedar (*Chamaecyparis*
    thyoides), *9*, *31–32*
  Baldcypress (*Taxodium distichum*), *9*,
    37–38
  Eastern redcedar (*Juniperus virginiana*
    var. virginiana), *9*, *33–34*
  guide to, *9*
  Pondcypress (*Taxodium ascendens*), *9*,
    35–36

Dahoon (*Ilex cassine*), margin of leaves,
  78
Darlington oak. *See* Laurel Oak
deciduous holly. *See* Possumhaw
dehiscent, 24–25, 324
deltate leaf shape, *7*
deltoid, 324
dentate leaf margin, *7*, 324
Devil's Walkingstick (*Aralia spinosa*),
  79–80

bark of large tree, *80*
flowers, *79*
fruit, *79–80*
leaf, *79*
winter twig, *310*
young stem, *80*
Devilwood (*Osmanthus americanus*),
26, 237–238
bark of young, *238*
flowers, *237*
leaves, *237*
leaves and fruit, *237*
diamond-leaf oak. *See* Swamp Laurel
Oak
diaphragmed pith, 324
dioecious, 324
Diospyros virginiana. See Common
Persimmon
divergent, 324
Dogwood Family (*Cornaceae*), 11, 13,
101–110
Alternate-Leaf Dogwood (*Cornus alternifolia*), 19, 101–102
Blackgum (*Nyssa sylvatica*), 19,
109–110
Flowering Dogwood (*Cornus florida*),
18–19, 103–104
guide to, 18–19
leaves alternate, 19
leaves opposite, 18–19
Swamp Tupelo (*Nyssa biflora*), 19,
107–108
Water Tupelo (*Nyssa aquatica*), 19,
105–106
doubly serrate leaf margin, 7, 15, 324
Downy Serviceberry (*Amelanchier arborea*), 27, 243–244
bark with cracks, *244*
bark with stripes, *244*
flowers, *243*
fruit, *243*
leaf, *243*
terminal bud, *319*
Drummond Maple (*Acer rubrum* var.
drummondi), 274

drupe, 324
Durand Oak (*Quercus durandii*), 23,
139–140
bark, *140*
leaves, *139*
Durand white oak. *See* Durand Oak
Dutch elm disease, 306, 308
dwarf hackberry. *See* Georgia
Hackberry
Dwarf Pawpaw (*Asimina parviflora*),
acuminate apex of leaves, *70*
dye bush. *See* Sweetleaf

Eastern Cottonwood (*Populus deltoides*),
28, 259–260
bark, *260*
fruit with cottonlike seeds, *259*
leaves, *259*
pistillate flowers, *259*
staminate flowers, *259*
winter twig, *319*
Eastern Hemlock (*Tsuga canadensis*),
9, 55–56
bark of young, *56*
needles and seed cones, *55*
smaller cones of compared to larger
cones of Carolina hemlock, *55*
very thick bark of large tree, *56*
eastern hophornbeam. *See*
Hophornbeam
eastern popular. *See* Eastern
Cottonwood
Eastern Redbud (*Cercis canadensis*), 19,
121–122
bark, *121–122*
flowers, *121*
fruit, *121*
leaf, *121*
winter twig, *312*
Eastern Redcedar (*Juniperus virginiana*
var. *virginiana*), 9
bark, *34*
form and habitat, *33*
leaves and seed cones (courtesy of Alan
Cressler), *34*

eastern red oak. *See* Northern Red Oak

Eastern Roughleaf Dogwood (*Cornus asperifolia*), 104

Eastern White Pine (*Pinus strobus*), 10, 49–50
  five needles per facicle, 49
  furrowed bark of large tree, 50
  needles, 49
  seed cones, 49
  smooth bark of young, 50

Ebenaceae. See Ebony Family

Ebony Family (*Ebenaceae*), 13, 111–112
  Common Persimmon (*Diospyros virginiana*), 111–112

Elliott, Stephen, 42

elliptical leaf shape, 7, 324

Elm Family (*Ulmaceae*), 15, 301–308
  American Elm (*Ulmus americana*), 30, 305–306
  guide to, 30
  Slippery Elm (*Ulmus rubra*), 30, 307–308
  Water-Elm (*Planera aquatica*), 30, 301–302
  Winged Elm (*Ulmus alata*), 30, 303–304

emarginate leaf apice, 8, 324

embedded, 324

emerald ash borer, 234, 236

empress-tree. *See* Royal Paulownia

entire leaf margin, 7, 324

Ericaceae. See Heath Family

erose, 324

Euphorbiaceae. See Spurge Family

evergreen magnolia. *See* Southern Magnolia

exserted, 324

Fabaceae. See Bean or Pea Family

Fagaceae. See Beech Family

Fagus grandifolia. See American Beech

falcate, 324

fall plum. *See* Mexican plum

false hickory. *See* Red Hickory

farkleberry. *See* Sparkleberry

fascicled needles, 3, 10, 324

"fatwood," 46

fetid buckeye. *See* Ohio Buckeye

fiddle oak. *See* Water Oak

Fig Family (*Moraceae*), 12, 14, 225–228
  guide to, 26
  Osage-Orange (*Maclura pomifera*), 26, 225–226
  Red Mulberry (*Morus rubra*), 26, 227–228

Figwort Family (*Scrophulariaceae*), 11, 285–286
  Royal Paulownia (*Paulownia tomentosa*), 285–286

filament, 324

fish-bait tree. *See* Southern Catalpa

flameleaf sumac. *See* Winged Sumac

Flatwoods Plum (*Prunus umbrellata*), 250

Florida anise. *See* Anise-Tree

Florida Maple (*Acer floridanum*), 29, 267–268
  bark, 268
  fruit, 267
  leaves, 267
  pistillate flowers, 267
  staminate flowers, 267
  winter twig, 320

flower, perfect, 326

flowering cornel. *See* Flowering Dogwood

Flowering Dogwood (*Cornus florida*), 103–104
  bark, 104
  flower, 103
  fruit, 103
  leaves, 103
  winter twig, 311

flowers, pistillate
  American Holly, 75
  Blackgum, 109–110
  Black Willow, 261
  Boxelder, 271
  Eastern Cottonwood, 259
  Florida Maple, 267

Hophornbeam, *89*
Hornbeam, *87–88*
Osage-Orange, *225*
Red Maple, *273*
Southern Bayberry, *229*
Sycamore, *239*
Water Oak, *163*
White Ash, *233*
flowers, staminate
  American Holly, *75*
  Blackgum, *109–110*
  Boxelder, *271*
  Eastern Cottonwood, *259*
  Florida Maple, *267*
  Green Ash, *235*
  Hophornbeam, *89*
  Hornbeam, *87–88*
  Mockernut Hickory, *199*
  Nutmeg Hickory, *191*
  Osage-Orange, *225*
  Red Maple, *273*
  Southern Bayberry, *229*
  Willow Oak, *167*
follicle, *325*
forest associations and forest types,
 Alabama, 1
forest tent caterpillar, *256*
form
  Eastern Redcedar, *33*
  Live Oak, *180*
  open-growth, Pecan, *188*
  Smooth Sumac, *65–66*
  Sourwood, *114*
  Water Tupelo, *106*
four-ranked, *325*
fragrant-olive. *See* Devilwood
Frangula caroliniana. See Carolina
  Buckthorn
Fraxinus americana. See White Ash
Fraxinus caroliniana. See Carolina Ash
Fraxinus pennsylvanica. See Green Ash
Fraxinus profunda. See Pumpkin Ash
Fraxinus quadrangulata. See Blue Ash
Fringe-Tree (*Chionanthus virginicus*),
  26–27, 231–232

bark, *232*
leaves and immature fruit, *231*
leaves showing purplish base of the
  petiole, *231*
winter twig, *318*
young leaves and flowers, *231*
fruit
  with cottonlike seeds, Black Willow, *261*
  with cottonlike seeds, Eastern Cotton-
    wood, *259*
  developing, Allegheny Chinkapin, *129*
  with grooves at nut apex, Scarlet Oak,
    137
  immature, Fringe-Tree, *231*
  immature, Georgia Hackberry, *99*
  immature, Gum Bumelia, *279*
  immature, Mexican Plum, *253*
  immature, Sparkleberry, *115*
  with knobby cap of acorn, White Oak,
    133
  Ogeechee Tupelo, *105*
  in pairs, Swamp Tupelo, *107*
  spiny, Ohio Buckeye, *265*
  with stinkbug on it, Cucumbertree, *211*
  and twig, Carolina Buckthorn, *241*
  and twig, Slippery Elm, *307*
  and twig, Sweetleaf, *297*
  velvety, Staghorn Sumac, *66*
  in winter, Chinaberry, *224*
furrowed bark
  of Cherrybark Oak, *166*
  defined, *325*
  of large Eastern White Pine, *50*
  of large Spruce Pine, *44*

Georgia Hackberry (*Celtis tenuifolia*),
  18, 99–100
  bark, *100*
  corky warts, *100*
  flowers, *99*
  leaves and immature fruit, *99*
  winter twig, *311*
Ginseng Family (*Araliaceae*), 16, 79–80
  Devil's Walkingstick (*Aralia spinosa*),
    79–80

glabrous, 325
glandular, 325
glaucous, 325
Gleditisia aguatica. See Waterlocust
Gleditisia triacanthos. See Honeylocust
Gleditsch, Johann, 124
globose, 325
Gordon, James, 300
gordonia. *See* Loblolly-Bay
Gordonia lasianthus. See Loblolly-Bay
grandsy gray-beard. *See* Fringe-Tree
great laurel magnolia. *See* Southern
 Magnolia
Green Ash (*Fraxinus pennsylvanica*), 27,
 235–236
 bark, *236*
 fruit, *235*
 leaflets, *235*
 staminate flowers, *235*
 winter twig, *318*
gum bully. *See* Gum Bumelia
Gum Bumelia (*Sideroxylon lanuginosum*), 29, 279–280
 bark of large tree, *280*
 bark of small tree, *280*
 flowers, *279*
 immature fruit, *279*
 leaves and twig with rusty hair, *279*
Gymnosperms
 defined, 325
 guide to, 9
 identification of using leaves and
  cones, 3
 See also Cypress Family; Pine Family

Hackberry (*Celtis occidentalis*), 18,
 97–98
 bark, *98*
 flowers, *97*
 fruit, *97*
 leaves, *97*
 winter twig, *311*
hairy sweetshrub. *See* Sweetshrub
Hales, Stephen, 292
Halesia diptera. See Two-Wing

Silverbell
Halesia tetraptera. See Carolina
 Silverbell
hammock, 325
hard elm. *See* Winged Elm
hard maple. *See* Sugar Maple
haws. *See* Hawthorns
Hawthorns (*Crataegus* spp.), 28,
 245–246
 bark, *246*
 flowers and leaves, *245*
 fruit and leaves, *245*
Hazel Alder (*Alnus serrulata*), 81–82
 bark, *82*
 flowers, *81*
 leaves and fruit, 18, *81*, *113*
 unique stringy yellow petals of
  Witch-Hazel flowers for comparison,
  82
 wavy margin and lopsided base of
  Witch-Hazel leaves for comparison,
  82
 winter twig, *310*
head, 325
heart pine. *See* Longleaf Pine
Heath Family (*Ericaceae*), 14, 15,
 113–116
 Sourwood (*Oxydendrum arboreum*),
  113–114
hedge-apple. *See* Osage-Orange
Hercules'-Club (*Zanthoxylum
 clava-herculis*)
 bark, *258*
 flowers, *257*
 fruit, *257*
 leaf, *257*
 winter twig, *319*
 See also Devil's Walkingstick
Hibiscus or Mallow Family (*Malvaceae*), 14, 221–222
 Basswood (*Tilia americana*), 221–222
Holly Family (*Aquifoliaceae*), 14, 15,
 71–78
 American Holly (*Ilex opaca*), 17, 75–76
 guide to, 17

Large Gallberry (*Ilex coriacea*), 17, 71–72
  leaves deciduous, 17
  leaves evergreen, 17
  Possumhaw (*Ilex decidua*), 73–74
  Yaupon (*Ilex vomitoria*), 17, 77–78
Honeylocust (*Gleditisia triacanthos*), 19, 123–124
  bark of small honeylocust with scaly plates, *124*
  bark of small honeylocust with thorns, *124*
  flowers, *123*
  fruit, *123*
  leaves, *123*
  winter twig, *312*
Hophornbeam (*Ostrya virginiana*), 18, 89–90
  bark, *90*
  fruit, *89*
  staminate and pistillate flowers, *89*
  winter twig, *310*
Hornbeam (*Carpinus caroliniana*), 18, 87–88
  bark, *88*
  fruit, 87–88
  leaves, *87*
  pistillate flowers, 87–88
  winter twig, *310*
horse-apple. *See* Osage-Orange
horse-sugar. *See* Sweetleaf

Ilex ambigua. *See* Carolina Holly
Ilex amelanchier. *See* Sarvis Holly
Ilex cassine. *See* Dahoon
Ilex coriacea. *See* Large Gallberry
Ilex decidua. *See* Possumhaw
Ilex glabra. *See* Inkberry
Ilex myrtifolia. *See* Myrtle-Leaved Holly
Ilex opaca. *See* American Holly
Ilex vomitoria. *See* Yaupon
Illicium floridanum. *See* Anise-Tree
imbricate
  bud scales, 5
  defined, 325

imperfect flower, 325
indehiscent, 23–24, 325
Indiana banana. *See* Pawpaw
Indian-bean. *See* Southern Catalpa
Indian cherry. *See* Carolina Buckthorn
inequilateral
  defined, 325
  leaf base, 8
inflorescence, 325
Inkberry (*Ilex glabra*), margin of leaves, 72
iron oak. *See* Post Oak
ironwood. *See* Hophornbeam; Hornbeam

jack oak. *See* Blackjack Oak
Judas-tree. *See* Eastern Redbud
Juglandaceae. *See* Walnut Family
Juglans cinerea. *See* Butternut
Juglans nigra. *See* Black Walnut
juneberry. *See* Downy Serviceberry
Juniperus virginiana var. *silicola*. *See* Southern Redcedar
Juniperus virginiana var. *virginiana*. *See* Eastern Redcedar
juvenile versus mature bark, 4

keeled, 325
kingnut hickory. *See* Shellbark Hickory
"knees"
  of Baldcypress, 37
  of Pondcypress, *36*

lanceolate leaf shape, 7, *78*, 325
Large Gallberry (*Ilex coriacea*)
  bark, 71–72
  fruit, 17, *71*
  leaves with spiny teeth above the middle, 17, *71*
  margin of Inkberry leaves for comparison, *72*
large-leaved cucumbertree. *See* Bigleaf Magnolia
lateral bud, 5, *319*, 325
Lauraceae. *See* Laurel Family

laurel cherry. *See* Cherry Laurel

Laurel Family (*Lauraceae*), 12, 205–208
  guide to, 25
  Redbay (*Persea borbonia*), 25, 205–206
  Sassafras (*Sassafras albidum*), 25,
    207–208

Laurel Oak (*Quercus hemisphaerica*), 21,
  143–144
  bark of large tree, *144*
  bark of young tree, *144*
  fruit, *143*
  leaves, *143*
  seedling with lobed leaves, *144*
  Shingle Oak leaves, longer with a pu-
    bescent to tomentose underside, for
    comparison, *144*
  See also Swamp Laurel Oak

leaf arrangement
  alternate, 3–4, 12–17, 19, *101*, 323
  opposite, 3–4, 11, 18–19, 326
  whorled, 3–4

leaflets
  ciliate leaflet margin, *195*
  Green Ash, *235*
  leaves with five and seven, *185, 197*
  leaves with five to nine, 24
  leaves with nine (sometimes seven) or
    more, 24–25
  leaves with 8 to 24, 23–24
  margin teeth at base of, *287*
  obovate, *126*

leaf scar, 4–5, 325

leaf spot, 112

leaves
  apices, *8*
  bases, *8*
  bipinnately compound, 4, 16–17, 323
  characteristics, 4
  compound, 11, 15–17, 324
  deciduous, 13–15, 17, 26–28
  evergreen, 12–14, 25, 27
  lobed, 12, 20–22, *135*, *144*
  margins (*See* margin, leaf)
  needlelike, 3
  palmately compound, 4, 28, 326

pinnately compound, *4*, 27, 28–29,
  326

scalelike, 3

shapes, *7*

simple, 11–15, 26–27, 29, 327

simple or compound, 4

sizes, 4

trifoliately compound, 328

tripinnately compound, 328

unlobed, 12–15, 21–23, *135*

  See also Angiosperms

legume, 325

lenticel, 5, 325

lily-of-the-valley tree. *See* Sourwood

linear leaf shape, *7*, 325

Liquidambar styraciflua. *See* Sweetgum

Liriodendron tulipifera. *See* Tulip-Poplar

littleleaf disease, 39

little shellbark hickory. *See* Shellbark
  Hickory

Live Oak (*Quercus virginiana*), 23
  bark, *179–180*
  form (courtesy of Nancy Loewenstein),
    180
  fruit, *179*
  leaves, *179*
  leaves, some with margin teeth, *179*

lobe
  defined, 325
  See also leaves, lobed

Loblolly-Bay (*Gordonia lasianthus*),
  299–300
  bark, *300*
  flower, *299*
  fruit, *299*
  leaf, *299*

Loblolly Pine (*Pinus taeda*), 10, 51–52
  bark of young, *52*
  needles with pollen and seed cones, *51*
  seed cones, *51*

loblolly pine forests, most dominant
  forest type in Alabama, 1

Longleaf Pine (*Pinus palustris*), 10,
  45–46
  bark, *46*

grass-stage seedling, *46*
large silver-white terminal buds, 45–46
needles, *45*
pollen cones, *46*
seed cone, *45*
longleaf pine forests, common forest
  type in Alabama, 1
long-pointed, 325
long straw pine. *See* Longleaf Pine
lustrous, 325

Maclura pomifera. See Osage-Orange
Maclure, W., 226
Magnol, P., 212, 214, 216, 218, 220
Magnolia acuminata. See
  Cucumbertree
Magnoliaceae. See Magnolia Family
Magnolia Family (*Magnoliaceae*), 12–13,
  209–220
  Bigleaf Magnolia (*Magnolia macro-
    phylla*), 26, 215–216
  Cucumbertree (*Magnolia acuminata*),
    26, 211–212
  guide to, 25–26
  leaves deciduous, 26
  leaves semi-evergreen or evergreen,
    25–26
  Southern Magnolia (*Magnolia grandi-
    flora*), 25–26, 213–214
  Sweetbay Magnolia (*Magnolia virgini-
    ana*), 25, 219–220
  Tulip-Poplar (*Liriodendron tulipifera*),
    26, 209–210
  Umbrella Magnolia (*Magnolia tripe-
    tala*), 26, 217–218
Magnolia grandiflora. See Southern
  Magnolia
Magnolia macrophylla. See Bigleaf
  Magnolia
Magnolia tripetala. See Umbrella
  Magnolia
Magnolia virginiana. See Sweetbay
  Magnolia
Mahogany Family (*Meliaceae*), 17,
  223–224

Chinaberry (*Melia azedarach*),
  223–224
Mallow Family. *See* Hibiscus or Mallow
  Family
Malus angustifolia. See Southern
  Crabapple
Malvaceae. See Hibiscus or Mallow
  Family
margin, leaf
  with coarse margin teeth, *292*
  crenate margin, *7*
  defined, 325
  dentate margin, *7, 324*
  doubly serrate margin, *7, 15, 324*
  entire margin, *7, 72, 324*
  irregularly toothed, 14–15, 29–30
  margin with callous-tipped teeth, *161*
  palmately lobed margin, *7*
  pinnately lobed margin, *7*
  regularly toothed, 29
  repand margin, *7*
  revolute margins, *146*
  serrate margin, *7, 14–15, 29, 58*
  sinuate margin, *7*
  with spiny teeth, *71*
  with teeth, 14–15, 29, *128, 179, 251*
  without teeth, 12–13, 29
  with teeth at base of leaflets, *287*
  wavy margin, *82*
marsh pine. *See* Pond Pine
meadow pine. *See* Loblolly Pine
Melia azedarach. See Chinaberry
Meliaceae. See Mahogany Family
mesic, 325
Mexican Plum (*Prunus mexicana*), 28
  bark, *254*
  flowers, *253*
  immature fruit, *253*
  leaves, *253*
  winter twig, *319*
midrib, 325
Mimosa (*Albizia julibrissin*), 19–20,
  119–120
  bark, *120*
  flowers, *119*

leaf, *119*
  leaves, flowers, and fruit, *119*
  winter twig, *312*
Mississippi hackberry. *See* Sugarberry
mixed loblolly pine-hardwood forests,
  common forest type in Alabama, 1
Mixed Mesophytic Association, 1
mixed upland hardwood forests, common forest type in Alabama, 1
Mockernut Hickory (*Carya tomentosa*),
  24, 199–200
  bark, *200*
  bark of young, *200*
  fruit, *199*
  leaf, *199*
  staminate flowers, *199*
  winter twig, *316*
mock-orange. *See* Osage-Orange
monoecious, 325
Moraceae. See Fig Family
Morella caroliniensis. See Swamp
  Candleberry
Morella cerifera. See Southern Bayberry
Morella inodora. See Odorless Bayberry
Morus alba. See White Mulberry
Morus rubra. See Red Mulberry
Moscatel Family (*Adoxaceae*), 17, 57–58
  Rusty Blackhaw (*Viburnum rufidulum*),
    57–58
mountain basswood. *See* Basswood
mountain magnolia. *See* Cucumbertree
mucilaginous, 326
mucronate
  defined, 326
  leaf apex, 8
musclewood. *See* Hornbeam
Myricaceae. See Wax-Myrtle Family
Myrtle-Leaved Holly (*Ilex myrtifolia*),
  lanceolate leaves, 78
Myrtle Oak (*Quercus myrtifolia*), revolute margins of leaves, 146

naked bud, 5, 326
narrow-leaved crabapple. *See* Southern
  Crabapple

needlelike leaves, 3
needles
  in bundles (fascicled), 3, 10, 324
  Longleaf Pine, 45
  with pollen and seed cones, Loblolly
    Pine, 51
  with pollen and seed cones, Virginia
    Pine, 53
  Pond Pine, 47
  and seed cones, Slash Pine, 41
  Shortleaf Pine, 39
  Spruce Pine, 43
  unbundled, 3, 9
node, 5, 326
Northern Catalpa (*Catalpa speciosa*),
  flowers, 92
northern hackberry. *See* Hackberry
northern hemlock. *See* Eastern
  Hemlock
Northern Red Oak (*Quercus rubra*), 21,
  169–170
  bark of large tree, *169–170*
  bark with white "ski tracks," *169–170*
  fruit, *169*
  leaves, *169*
  winter twig, *315*
northern white pine. *See* Eastern White
  Pine
nut
  Black Walnut, *203*
  defined, 326
  and leaves, Swamp Laurel Oak, 21–22
  smooth or ribbed, enclosed in a dehis-
    cent husk; pith solid, Walnut Family,
    24–25
  striped acorn, Pin Oak, *138*
  and thick husk, Shagbark Hickory, *195*
  and thick husk, Shellbark Hickory, *189*
nutlet, 326
Nutmeg Hickory (*Carya myristiciformis*), 25, 191–192
  bark of young, *192*
  fruit, *191*
  leaf and fruit, *191*
  staminate flowers, *191*

winter twig, *316*
Nuttall Oak (*Quercus texana*), 20–21,
175–176
  bark, *176*
  fruit, *175*
  leaf, *175*
  winter twig, *315*
Nyssa aquatica. See Water Tupelo
Nyssa biflora. See Swamp Tupelo
Nyssa ogeche. See Ogeechee Tupelo
Nyssa sylvatica. See Blackgum

Oak-Hickory Association, 1
oblanceolate, 326
oblong
  defined, 326
  leaf shape, 7
obovate
  defined, 326
  leaf, *161*
  leaflets, *126*
obtuse
  defined, 326
  leaf apice, *8*
  leaf base, *8*
Odorless Bayberry (*Morella inodora*),
230
Ogeechee Tupelo (*Nyssa ogeche*)
  leaves and fruit, 105–106
  winter twig, *311*
Ohio Buckeye (*Aesculus glabra*), 28
  bark, *266*
  flowers (courtesy of John Seiler), *265*
  leaf, *265*
  spiny fruit, *265*
  winter twig, *320*
oil nut. *See* Butternut
old field pine. *See* Loblolly Pine
old-man's beard. *See* Fringe-Tree
Oleaceae. See Olive Family
Olive Family (*Oleaceae*), 11, 231–238
  Devilwood (*Osmanthus americanus*),
  26–27, 237–238
  Fringe-Tree (*Chionanthus virginicus*),
  27, 231–232

Green Ash (*Fraxinus pennsylvanica*),
27, 235–236
guide to, 26–27
leaves pinnately compound, 27
leaves simple, 26–27
White Ash (*Fraxinus americana*), 27,
233–234
opossum oak. *See* Water Oak
opposite leaf arrangement, 3–4, 11,
18–19, 326
orbicular
  defined, 326
  leaf shape, 7
Osage-Orange (*Maclura pomifera*), 1,
26, 225–226
  bark, *226*
  fruit, *226*
  leaves, *225*
  pistillate flowers, *225*
  winter twig, *318*
Osmanthus americanus. See Devilwood
Ostrya virginiana. See Hophornbeam
oval
  defined, 326
  leaf shape, 7
ovary, 326
ovate
  defined, 326
  leaf shape, 7
Overcup Oak (*Quercus lyrata*), 22,
151–152
  bark, *152*
  fruit, *151*
  larger Bur Oak acorn and cap for comparison, *152*
  leaves, *151*
  lobing and sinuses of Bur Oak leaves
  for comparison, *152*
  winter twig, *314*
ovoid, 326
ovule, 326
Oxydendrum arboretum. See Sourwood

Painted Buckeye (*Aesculus sylvatica*),
264

pale hickory. *See* Sand Hickory
pale-leaf hickory. *See* Sand Hickory
palmate
  defined, 326
  palmately compound leaves, 4, 28
  palmately lobed leaf margin, 7
panicle, 326
Paper-Mulberry (*Broussonetia papyrifera*), 228
Paulownia tomentosa. See Royal Paulownia
Pawpaw (*Asimina triloba*), 69–70
  acuminate apex of Dwarf Pawpaw leaves for comparison, 70
  bark, 69–70
  fruit, 69
  leaves, 69
  lowers, 69
  winter twig, 310
peach oak. *See* Willow Oak
Pecan (*Carya illinoinensis*), 25, 187–188
  bark, 188
  fruit, 187
  leaf, 187
  open-growth form, 188
  winter twig, 316
pecan hickory. *See* Bitternut Hickory; Pecan; Water Hickory
peltate, 326
pendent, 326
perfect flower, 326
Persea borbonia. See Redbay
Persea palustris. See Swamp Redbay
petal, 326
petiole
  defined, 326
  of Ogeechee Tupelo leaves, 105
  purplish base of, Fringe-Tree, 231
  of Smooth Blackhaw leaves, 58
phloem, 5
Pignut Hickory (*Carya glabra*), 24, 185–186, 193–194
  bark, 186–187
  fruit, 185
  leaf with five leaflets, 185

leaf with seven leaflets, 185
  and Red Hickory (*Carya ovalis*), 193–194
  winter twig, 316
Pinaceae. See Pine Family
Pine Family (*Pinaceae*), 39–56
  Eastern Hemlock (*Tsuga canadensis*), 9, 55–56
  Eastern White Pine (*Pinus strobus*), 10, 49–50
  guide to, 9–10
  Loblolly Pine (*Pinus taeda*), 10, 51–52
  Longleaf Pine (*Pinus palustris*), 10, 45–46
  needles in bundles of more than two, 10
  needles in bundles of two and three, 10
  needles not in bundles, 9
  needles only in bundles ot two, 10
  Pond Pine (*Pinus serotina*), 10, 47–48
  Shortleaf Pine (*Pinus echinata*), 10, 39–40
  Slash Pine (*Pinus elliottii*), 10, 41–42
  Spruce Pine (*Pinus glabra*), 10, 43–44
  Virginia Pine (*Pinus virginiana*), 10, 53–54
pinnately compound leaves, 4, 27, 28–29, 326
pinnately lobed leaf margin, 7
Pin Oak (*Quercus palustris*), leaf and striped acorn of, 138
Pinus clausa. See Sand Pine
Pinus echinata. See Shortleaf Pine
Pinus elliottii. See Slash Pine
Pinus glabra. See Spruce Pine
Pinus palustris. See Longleaf Pine
Pinus rigida. See Pitch Pine
Pinus serotina. See Pond Pine
Pinus strobus. See Eastern White Pine
Pinus taeda. See Loblolly Pine
Pinus virginiana. See Virginia Pine
pistil, 326
pistillate
  defined, 326

See also flowers, pistillate
Pitch Pine (*Pinus rigida*), 48
pith, 5
  chambered, 324
  defined, 326
  diaphragmed, 324
Planer, Johann, 302
Planera aquatica. *See* Water-Elm
planertree. *See* Water-Elm
Platanaceae. *See* Sycamore Family
Platanus occidentalis. *See* Sycamore
pocosin, 326
pocosin pine. *See* Pond Pine
poison-elderberry. *See* Poison-Sumac
Poison-Sumac (*Toxicodendron vernix*),
  67–68
  bark, 68
  fruit, 67
  leaves, 17, 67
  winter twig, 309
polygamous, 326
pome, 327
Pondcypress (*Taxodium ascendens*), 9,
  35–36
  bark, 36
  bark and "knees," 36
  leaves, 35
  seed cone, 35
Pond Pine (*Pinus serotina*), 10, 47–48
  bark, 48
  bark with sprouts, 48
  needles, 47
  serotinous seed cones, 47
popcorn tree. *See* Chinese Tallowtree
Populus deltoides. *See* Eastern
  Cottonwood
Populus heterophylla. *See* Swamp
  Cottonwood
Possumhaw (*Ilex decidua*), 73–74
  bark, 73–74
  fruit, 73
  leaves and flowers, 73
  leaves and fruit, 17
Possumhaw (*Viburnum nudum*), 58
possumwood. *See* Common

Persimmon
Post Oak (*Quercus stellata*), 22
  bark of large tree, 174
  fruit, 173
  leaf with cruciform central lobes, 173
  leaves with variable lobing, 173
  winter twig, 315
prickle, 327
prickly-ash. *See* Devil's Walkingstick
Pride of India. *See* Chinaberry
princess-tree. *See* Royal Paulownia
Prunus alabamensis. *See* Alabama Black
  Cherry or Alabama Chokeberry
Prunus americana. *See* American Plum
Prunus angustifolia. *See* Chickasaw
  Plum
Prunus caroliniana. *See* Cherry Laurel
Prunus mexicana. *See* Mexican Plum
Prunus serotina. *See* Black Cherry
Prunus umbrellata. *See* Flatwoods Plum
pubescent
  defined, 327
  Shingle Oak leaves, 144
Pumpkin Ash (*Fraxinus profunda*), 236

Quassia Family (*Simaroubaceae*), 16,
  287–288
  tree-of-heaven (*Ailanthus altissima*),
  287–288
quercitron oak. *See* Black Oak
Quercus alba. *See* White Oak
Quercus arkansana. *See* Arkansas Oak
Quercus austrina. *See* Bluff Oak
Quercus coccinea. *See* Scarlet Oak
Quercus durandii. *See* Durand Oak
Quercus falcata. *See* Southern Red Oak
Quercus geminata. *See* Sand Live Oak
Quercus hemisphaerica. *See* Laurel Oak
Quercus imbricaria. *See* Shingle Oak
Quercus incana. *See* Bluejack Oak
Quercus laevis. *See* Turkey Oak
Quercus laurifolia. *See* Swamp Laurel
  Oak
Quercus lyrata. *See* Overcup Oak
Quercus macrocarpa. *See* Bur Oak

Quercus margarettiae. See Sand Post Oak

Quercus marilandica. See Blackjack Oak

Quercus michauxii. See Swamp Chestnut Oak

Quercus montana. See Chestnut Oak

Quercus muehlenbergii. See Chinkapin Oak

Quercus myrtifolia. See Myrtle Oak

Quercus nigra. See Water Oak

Quercus pagoda. See Cherrybark Oak

Quercus palustris. See Pin Oak

Quercus phellos. See Willow Oak

Quercus rubra. See Northern Red Oak

Quercus shumardii. See Shumard Oak

Quercus stellata. See Post Oak

Quercus texana. See Nuttall Oak

Quercus velutina. See Black Oak

Quercus virginiana. See Live Oak

raceme, 327

rachis, 327

recurved, 327

Redbay (*Persea borbonia*), 25, 205–206

  bark of young, *206*

  fruit, *205*

  leaves, *205*

  leaves showing insect galls, *205*

red birch. *See* River Birch

Red Buckeye (*Aesculus pavia*), 264

redbud. *See* Eastern Redbud

red elm. *See* Slippery Elm

red gum. *See* Sweetgum

Red Hickory (*Carya ovalis*), 24, 193–194

  bark, *194*

  fruit, *193*

  leaf and fruit, *193*

red juniper. *See* Eastern Redcedar

Red Maple (*Acer rubrum* var. *rubrum*), 29, 273–274

  bark, *274*

  fruit, *273*

  pistillate flowers, *273*

  staminate flowers, *273*

  winter twig, *320*

Red Mulberry (*Morus rubra*), 26, 227–228

  bark, *228*

  fruit, *227*

  leaf, *227*

  leaves with lobing, *227*

  staminate and pistillate flowers, *227*

  winter twig, *318*

red oak. *See* Cherrybark Oak; Northern Red Oak; Scarlet Oak; Shumard oak

reflexed, 327

reniform leaf shape, *7*

repand

  defined, 327

  leaf margin, *7*

revolute, 327

Rhamnaceae. See Buckthorn Family

rhombic

  defined, 327

  leaf shape, *7*

Rhus copallinum. See Winged Sumac

Rhus glabra. See Smooth Sumac

Rhus typhina. See Staghorn Sumac

River Birch (*Betula nigra*), 18, 85–86

  flowers, *85*

  fruit, *85*

  leaves, *85*

  peeling bark of, *86*

  scaly bark of, *86*

  winter twig, *310*

Robin, Jean, 126

Robinia pseudoacacia. See Black Locust

rock oak. *See* Chestnut Oak

Rosaceae. See Rose Family

Rose Family (*Rosaceae*), 12, 14, 15, 243–256

  Black Cherry (*Prunus serotina*), 28, 255–256

  Cherry Laurel (*Prunus caroliniana*), 27, 251–252

  Chickasaw Plum (*Prunus angustifolia*), 27–28, 249–250

  Downy Serviceberry (*Amelanchier arborea*), 27, 243–244

  guide to, 27–28

Hawthorns (*Crataegus* spp.), 28,
  245–246
leaves deciduous, 27–28
leaves evergreen, 27
Mexican Plum (*Prunus mexicana*), 28,
  253–254
Southern Crabapple (*Malus angustifo-*
  lia), 27, 247–248
rosemary pine. *See* Shortleaf Pine
rounded leaf apice, 8
rounded leaf base, 8
Royal Paulownia (*Paulownia tomen-*
  tosa), 285–286
  bark, *286*
  flowers, *285*
  fruit, *285*
  leaf, *285*
  winter twig, *321*
Rue Family (*Rutaceae*), 15–16, 257–258
  Hercules'-Club (*Zanthoxylum cla-*
  va-herculis), 257–258
rugose, 327
rum cherry. *See* Black Cherry
Rusty Blackhaw (*Viburnum rufidulum*),
  57–58
  bark, *58*
  fruit (courtesy of Nancy Loewenstein),
  57
  leaves, *57*
  leaves and flowers, *57*
  margin of Southern Arrowwood leaves
  for comparison, *58*
  Smooth Blackhaw leaves for comparison, *58*
  winter twig, *309*
rusty nannyberry. *See* Rusty Blackhaw
Rutaceae. See Rue Family

Salicaceae. See Willow or Poplar Family
Salix nigra. See Black Willow
samara, 327
Sand Hickory (*Carya pallida*), 24,
  197–198
  bark, *198*
  fruit, *197*

leaf, with five leaflets and pale underside, and fruit, *197*
leaf with seven leaflets, *197*
winter twig, *316*
sand holly. *See* Carolina Holly
sand jack oak. *See* Bluejack Oak
sand laurel oak. *See* Laurel Oak
Sand Live Oak (*Quercus geminata*), 180
Sand Pine (*Pinus clausa*), orange-gray
  scaly bark, *40*
Sand Post Oak (*Quercus margarettiae*),
  22, 153–154
  bark, *154*
  fruit, *153*
  leaves, *153*
  leaves showing variation in lobing, *154*
  winter twig, *314*
Sapindaceae. See Soapberry Family
Sapodilla Family (*Sapotaceae*), 12,
  279–282
  Buckthorn Bumelia (*Sideroxylon lycioi-*
  des), 29, 281–282
  guide to, 29
  Gum Bumelia (*Sideroxylon lanugino-*
  sum), 29, 279–280
Sapotaceae. See Sapodilla Family
Sarvis Holly (*Ilex amelanchier*), 74
Sassafras (*Sassafras albidum*), 25
  bark, *208*
  in flower, *207*
  fruit, *207*
  leaves, *207*
  winter twig, *317*
Sassafras albidum. See Sassafras
satin walnut. *See* Sweetgum
scabrous, 327
scalelike leaves, 3
scaly bark
  of Black Cherry, *256*
  of Carolina Silverbell, *292*
  orange-gray, of Sand Pine, *40*
  plates, of small Honeylocust, *124*
  of River Birch, *86*
  of Sycamore, *240*
scalybark hickory. *See* Shagbark Hickory

Scarlet Oak (*Quercus coccinea*), 21, 137–138
  bark with flat ridges, 138
  bark with white streaks, 138
  fruit (note the grooves at the nut apex), 137
  leaf and striped acorn of Pin Oak for comparison, 138
  shade leaf, 137
  winter twig, 313
scarlet sumac. *See* Smooth Sumac
Schisandraceae. *See* Star-Vine Family
scientific naming system, 1–2
Scrophulariaceae. *See* Figwort Family
scrubby post oak. *See* Sand Post Oak
scrub chestnut oak. *See* Chinkapin Oak
scrub oak. *See* Bluejack Oak; Turkey oak
scrub pine. *See* Virginia Pine
scurfy, 327
seed cones
  Atlantic White-Cedar, 31
  Eastern Hemlock, 55
  Eastern Redcedar, 34
  Eastern White Pine, 49
  Loblolly Pine, 51
  serotinous, Pond Pine, 47
  Shortleaf Pine, 39
  Slash Pine, 41
  Spruce Pine, 43
  Virginia Pine, 53
sepal, 327
September Elm (*Ulmus serotina*), 304
serotinous
  defined, 327
  seed cones, 47
serrate leaf margin, 7, 14–15, 29, 58
sessile, 327
shadbush. *See* Downy Serviceberry
shade leaf
  defined, 327
  of Scarlet Oak, 137
shade tolerance, 327
Shagbark Hickory (*Carya ovata*), 24, 195–196
  bark, 196
  ciliate leaflet margin, 195
  fruit, 195
  leaf, 195
  nut and thick husk, 195
  winter twig, 316
Shellbark Hickory (*Carya laciniosa*), 24, 189–190
  bark, 190
  fruit, 189
  leaf, 189
  nut and thick husk, 189
  winter twig, 316
Shingle Oak (*Quercus imbricaria*), leaves, with a pubescent to tomentose underside, 144
shining sumac. *See* Winged Sumac
Shortleaf Pine (*Pinus echinata*), 10, 39–40
  bark, 39–40
  needles, 39
  orange-gray scaly bark of sand pine for comparison, 40
  with resin holes in the bark, 40
  seed cones, 39
short straw pine. *See* Shortleaf Pine
Shumard Oak (*Quercus shumardii*), 21, 171–172
  bark, 172
  fruit, 171
  leaf, 171
  winter twig, 315
Sideroxylon lanuginosum. *See* Gum Bumelia
Sideroxylon lycioides. *See* Buckthorn Bumelia
silk tree. *See* Mimosa
Silky Dogwood (*Cornus amomum*), 104
Silver Maple (*Acer saccharinum*), 29, 275–276
  bark, 276
  flowers, 275
  fruit, 275
  leaf, 275
  winter twig, 320

Silvics of North America, volumes 1 and 2
(Burns and Honkala, eds.), 2
Simaroubaceae. See Quassia Family
simmon. *See* Common Persimmon
simple
defined, 327
See also leaves, simple
sinuate leaf margin, 7, 327
sinus, 327
"ski tracks," white, in bark of Northern
Red Oak, 169–170
Slash Pine (*Pinus elliottii*), 10, 41–42
bark with brown flecking, 42
bark with pale flecking, 42
needles and seed cones, 41
slash pine forests, common forest type
in Alabama, 1
Slippery Elm (*Ulmus rubra*), 30,
307–308
bark, 308
flowers, 307
fruit and twig, 307
leaves, 307
small-leaved elm. *See* Winged Elm
Smoketree (*Cotinus obovatus*), winter
twig, 309
Smooth Blackhaw (*Viburnum prunifolium*), shape and petiole of leaves, 58
smooth hickory. *See* Pignut Hickory
Smooth Sumac (*Rhus glabra*), 17, 65–66
bark, 66
flowers, 65
form, 65–66
fruit, 65
leaves, 17, 65
velvety fruit of Staghorn Sumac for
comparison, 66
winter twig, 309
snowdrop tree. *See* Carolina Silverbell
Soapberry Family (*Sapindaceae*), 11,
263–278
Boxelder (Acer negundo), 28–29,
271–272
Chalk Maple (*Acer leucoderme*), 29,
269–270

Florida Maple (*Acer floridanum*), 29,
267–268
guide to, 28–29
leaves palmately compound, 28
leaves pinnately compound, 28–29
leaves simple
margins not toothed, 29
margins serrate, 29
Ohio Buckeye (*Aesculus glabra*), 28,
265–266
Red Maple (*Acer rubrum* var. *rubrum*),
29, 273–274
Silver Maple (*Acer saccharinum*), 29,
275–276
Sugar Maple (*Acer saccharum*), 29,
277–278
Yellow Buckeye (*Aesculus flava*), 28,
263–264
soft elm. *See* Slippery Elm
soft maple. *See* Red Maple; Silver Maple
sorrel tree. *See* Sourwood
sourgum. *See* Blackgum
Sourwood (*Oxydendrum arboretum*),
113–114
bark, 114
bark with orange in the grooves, 114
flowers, 113
form and habitat, 114
fruit, 113
green twig, 312
leaves, 113
red twig, 312
sourwood honey, 114
Southern Arrowwood (*Viburnum dentatum*), leaf margin, 58
southern basswood. *See* Basswood
Southern Bayberry (*Morella cerifera*),
229–230
bark, 230
fruit, 229
leaves, 229
pistillate and staminate flowers, 229
Southern Catalpa (*Catalpa bignonioides*), 91–92
bark, 92

flowers, *91*
fruit, *91*
leaf, *91*
Northern Catalpa flowers, *92*
whorled leaf arrangement, 4
winter twig, *310*
southern cottonwood. *See* Eastern
Cottonwood
Southern Crabapple (*Malus angustifolia*), 27, 247–248
bark, *248*
flower, *247*
fruit, *247*
leaves, *247*
winter twig, *319*
southern cypress. *See* Baldcypress
southern hackberry. *See* Sugarberry
Southern Magnolia (*Magnolia grandiflora*), 25, 213–214
flower, *213*
fruit, *213*
leaves, *213*
smooth bark, *214*
winter twig, *317*
Southern Mixed Hardwoods Association, 1
southern prickly-ash. *See*
Hercules'-Club
Southern Redcedar (*Juniperus virginiana* var. *silicola*), 34
Southern Red Oak (*Quercus falcata*),
20, 141–142
bark, *142*
fruit, *141*
leaves, *141*
leaves of seedlings, *142*
winter twig, *313, 314*
Southern Shagbark Hickory (*Carya
carolinae-septentrionalis*), 196
southern sugar maple. *See* Florida
Maple
southern white-cedar. *See* Atlantic
White-Cedar
southern yellow pine. *See* Longleaf
Pine; Shortleaf Pine; Slash Pine

Spanish oak. *See* Scarlet Oak; Southern
Red Oak
Sparkleberry (*Vaccinium arboreum*),
115–116
bark, *116*
flowers, *115*
immature fruit, *115*
leaves, *115*
spatulate leaf shape, 7
spine, 327
Spruce Pine (*Pinus glabra*), 10, 43–44
furrowed bark on large tree, *44*
needles, *43*
seed cones, *43*
smooth or shallowly ridged bark, *44*
Spurge Family (*Euphorbiaceae*), 13,
117–118
Chinese Tallowtree (*Triadica sebifera*),
117–118
Staghorn Sumac (*Rhus typhina*), velvety
fruit of, 66
stalked, 327
stamen, 327
Staphyleaceae. *See* Bladdernut Family
Staphylea trifolia. *See* American
Bladdernut
Star-Vine Family (*Schisandraceae*), 12,
283–284
Anise-Tree (*Illicium floridanum*),
283–284
stave oak. *See* White Oak
stellate, 328
stigma, 328
stink-bush. *See* Anise-Tree
stinking buckeye. *See* Ohio Buckeye
stipular scar, 328
stipule, 328
Storax Family (*Styracaceae*), 13, 15,
291–296
American Snowbell (*Styrax americanus*), 29–30, 293–294
Bigleaf Snowbell (*Styrax grandifolius*),
30, 295–296
Carolina Silverbell (*Halesia tetraptera*),
29, 291–296

guide to, 29–30
leaf margin irregularly toothed, 29–30
leaf margin regularly toothed, 29
strawberry-bush. *See* Sweetshrub
Strawberry-Shrub Family (*Calycantha-ceae*), 11, 93–94
 Sweetshrub (*Calycanthus floridus*), 93–94
striations, 328
striped bark
 of Carolina Silverbell, *292*
 of Downy Serviceberry, *244*
striped oak. *See* Nuttall Oak
style, 328
Styracaceae. See Storax Family
Styrax americanus. See American Snowbell
Styrax grandifolius. See Bigleaf Snowbell
subopposite, 328
Sugarberry (*Celtis laevigata*), 18, 95–100
 bark, 95–96
 flowers, 95
 fruit, 95
 leaves, 95
 winter twig, *311*
sugar hackberry. *See* Sugarberry
Sugar Maple (*Acer saccharum*), 29
 bark, *278*
 flowers, *277*
 fruit, *277*
 leaves, *277*
 winter twig, *320*
sun leaf
 Black Oak, *177*
 defined, 328
superposed bud, 328
swamp, 328
swamp ash. *See* Green Ash
swamp bay. *See* Sweetbay Magnolia
swamp blackgum. *See* Swamp Tupelo
Swamp Candleberry (*Morella carolinien-sis*), 230
swamp cedar. *See* Atlantic White-Cedar
Swamp Chestnut Oak (*Quercus*

michauxii), 23, 157–158
 bark, *158*
 fruit, *157*
 leaves, *157*
 winter twig, *314*
Swamp Cottonwood (*Populus hetero-phylla*), 260
Swamp Dogwood (*Cornus foemina*), 104
swamp hickory. *See* Bitternut Hickory; Nutmeg Hickory
Swamp Laurel Oak (*Quercus laurifolia*), 149–150
 bark, *150*
 fruit, *149*
 leaves, *149*
 leaves and nut, 21–22
 leaves showing subrhombic shape, *150*
swamp magnolia. *See* Sweetbay Magnolia
swamp pine. *See* Slash Pine
Swamp Redbay (*Persea palustris*), 206
swamp red oak. *See* Cherrybark Oak; Shumard Oak
swamp-sumac. *See* Poison-Sumac
Swamp Tupelo (*Nyssa biflora*), 19, 107–108
 bark, *108*
 base and bark, *108*
 fruit in pairs, *107*
 leaves, *107*
 winter twig, *311*
swamp willow. *See* Black Willow
swamp willow oak. *See* Willow Oak
Sweetbay Magnolia (*Magnolia virgini-ana*), 25, 219–220
 bark, *220*
 flower, *219*
 fruit, *219*
 silvery underside of leaves, *219*
sweetbay-swamp tupelo-red maple forests, common forest type in Ala-bama, 1
sweet birch. *See* Black Birch
sweet buckeye. *See* Yellow Buckeye
sweet gallberry. *See* Large gallberry

Sweetgum Family (*Altingiaceae*), 59–60
  Sweetgum (*Liquidambar styraciflua*), 59–60
Sweetgum (*Liquidambar styraciflua*), 12, 59–60
  bark, *60*
  flowers, *59*
  fruit, *59*
  leaf, *59*
  winter twig, *309*
sweetgum-Nutall oak-willow oak forests, common forest type in Alabama, 1
sweetgum-tulip-poplar forests, common forest type in Alabama, 1
Sweetleaf Family (*Symplocaceae*), 12, 297–298
  Sweetleaf (*Symplocos tinctoria*), 297–298
Sweetleaf (*Symplocos tinctoria*), 297–298
  bark, *298*
  flowers, *297*
  leaves and flower buds, *297*
sweet-locust. *See* Honeylocust
sweet maple. *See* Sugar Maple
sweet pecan. *See* Pecan
sweet pignut. *See* Pignut Hickory
Sweetshrub (*Calycanthus floridus*), 93–94
  flower, *93–94*
  fruit, *94*
  leaf underside, *93*
  leaves, *93*
  seedling, *94*
  young stem, *94*
Sycamore Family (*Platanaceae*), 239–240
  Sycamore (*Platanus occidentalis*), 239–240
Sycamore (*Platanus occidentalis*), 12, 239–240
  with exposed pale bark, *240*
  fruit, *239*
  leaves and stipules, *239*

pistillate flowers, *239*
scaly bark, *240*
winter twig, *318*
Symplocaceae. See Sweetleaf Family
Symplocos tinctoria. See Sweetleaf

tag alder. *See* Hazel Alder
tanbark oak. *See* Chestnut Oak
Taxodium ascendens. See Pondcypress
Taxodium distichum. See Baldcypress
Tea Family (*Theaceae*), 14, 299–300
  Loblolly-Bay (*Gordonia lasianthus*), 299–300
terminal bud, 5
  defined, 328
  Downy Serviceberry, *319*
  Longleaf Pine, 45–46
Texas red oak. *See* Nuttall Oak
Theaceae. See Tea Family
thorn, 328
thorny-locust. *See* Honeylocust
thunderwood. *See* Poison-Sumac
Tilia americana. See Basswood
tomentose
  defined, 328
  Shingle Oak leaves, *144*
toothache-tree. *See* Hercules'-Club
Toxicodendron vernix. See Poison-Sumac
tree huckleberry. *See* Sparkleberry
Tree-of-Heaven (*Ailanthus altissima*)
  bark, *288*
  flowers, *287*
  fruit, *287*
  leaf (note margin teeth at base of leaflets), *287*
  winter twig, *321*
trees
  how to identify, 3–5
  identification features, 7–8
Triadica sebifera. See Chinese Tallowtree
trifoliately compound, 328
tripinnately compound, 328
truncate

defined, 328
leaf apex, *8*
leaf base, *8*
Tsuga canadensis. See Eastern Hemlock
Tsuga carolinina. See Carolina Hemlock
Tulip-Poplar (*Liriodendron tulipifera*), 26, 209–210
  bark of large tree, *210*
  bark of young, *210*
  fruit, *209*
  leaves and flower, *209*
  winter twig, *317*
tuliptree. *See* Tulip-Poplar
tupelo gum. *See* Blackgum; Water Tupelo
tupelo honey, 110
Turkey Oak (*Quercus laevis*), 20, 147–148
  bark, *148*
  fruit, *147*
  leaf, *147*
  leaves with turkey track outline, *148*
  winter twig, *314*
two-flowered tupelo. *See* Swamp Tupelo
two-ranked, 328
Two-Wing Silverbell (*Halesia diptera*)
  coarse margin teeth of leaf, *292*
  petals fused only at the base of, *292*
  winter twig, *321*

Ulmaceae. See Elm Family
Ulmus alata. See Winged Elm
Ulmus americana. See American Elm
Ulmus rubra. See Slippery Elm
Ulmus serotina. See September Elm
umbel, 328
Umbrella Magnolia (*Magnolia tripetala*), 26, 217–218
  bark, *218*
  fruit, *218*
  leaves, underside of, *217*
  leaves (courtesy of Alan Cressler), *217*
  winter twig, *317*
umbrella tree. *See* Umbrella Magnolia
unbundled needles, 3, 9

United States Department of Agriculture (USDA)
  Forest Service Forest Inventory and Analysis National Program, 1
  Natural Resources Conservation Service PLANTS Database, 2

Vaccinium arboreum. See Sparkleberry
valvate
  bud scales, 5, 25
  defined, 328
vein axil, 328
Viburnum cassinoides. See Witherod
Viburnum dentatum. See Southern Arrowwood
Viburnum nudum. See Possumhaw
Viburnum prunifolium. See Smooth Blackhaw
Viburnum rufidulum. See Rusty Blackhaw
Virginia live oak. *See* Live Oak
Virginia Pine (*Pinus virginiana*), 10, 53–54
  bark of large tree, *54*
  bark of young, *54*
  needles with the pollen and seed cones, *53*
  seed cones, *53*

wahoo. *See* Winged Elm
Walnut Family (*Juglandaceae*), 16, 181–204
  Bitternut Hickory (*Carya cordiformis*), 25, 183–184
  Black Walnut (*Juglans nigra*), 24, 203–204
  Butternut (*Juglans cinerea*), 23–24, 201–202
  guide to, 23–25
  leaves with 8 to 24 leaflets; twig pith chambered; nut corrugated or rugose, enclosed in an indehiscent husk, 23–24
  Mockernut Hickory (*Carya tomentosa*), 24, 199–200

Nutmeg Hickory (*Carya myristici-formis*), 25, 191–192
nut smooth or ribbed, enclosed in a dehiscent husk; pith solid, 24–25
   leaves with five to nine leaflets; buds imbricate; fruit husk unwinged at sutures, 24
   leaves with nine (sometimes seven) or more leaflets; buds valvate; fruit husk winged at sutures, 25
Pecan (*Carya illinoinensis*), 25, 187–188
Pignut Hickory (*Carya glabra*), 24, 185–186
Red Hickory (*Carya ovalis*), 24, 193–194
Sand Hickory (*Carya pallida*), 24, 197–198
Shagbark Hickory (*Carya ovata*), 24, 195–196
Shellbark Hickory (*Carya laciniosa*), 24, 189–190
Water Hickory (*Carya aquatica*), 25, 181–182
Walter's pine. *See* Spruce Pine
water ash. *See* Green Ash
water birch. *See* River Birch
water elm. *See* American Elm; Water-Elm
Water-Elm (*Planera aquatica*), 30, 301–302
   bark, 302
   flowers, 301
   fruit, 301
   leaves, 301
   winter twig, 321
watergum. *See* Water Tupelo
Water Hickory (*Carya aquatica*), 25, 181–182
   bark, 181–182
   fruit, 181
   leaf, 181
   with very shaggy bark, 182
   winter twig, 315
Waterlocust (*Gleditisia aguatica*), 124
Water Oak (*Quercus nigra*), 22, 163–164

bark of large tree, 164
fruit, 163–164
leaves, 163
leaves of sapling, 163
pistillate flower, 163–164
winter twig, 314
Water Tupelo (*Nyssa aquatica*), 19
   bark, 106
   form and habitat, 106
   fruit, 105
   leaves, 105
   shorter petioles of Ogeechee tupelo leaves and its fruit for comparison, 105
wax-myrtle. *See* Southern Bayberry
Wax-Myrtle Family (*Myricaceae*), 12, 229–230
   Southern Bayberry (*Morella cerifera*), 229–230
White Ash (*Fraxinus americana*), 27, 233–234
   bark, 234
   fruit, 233
   leaf, 233
   leaf scar, 233
   pistillate flowers, 233
white-bark maple. *See* Chalk Maple
white basswood. *See* Basswood
white bay. *See* Sweetbay Magnolia
white elm. *See* American Elm
white hickory. *See* Mockernut Hickory; Pignut Hickory
white holly. *See* American Holly
white maple. *See* Silver Maple
White Mulberry (*Morus alba*), 228
White Oak (*Quercus alba*), 22, 133–134
   bark with loose plates, 134
   flowers, 133
   fruit (note the knobby cap of the acorn), 133
   grooved bark, 134
   leaf, 133
   winter twig, 313
white oak-red oak-hickory forests, common forest type in Alabama, 1

white pine. *See* Eastern White Pine
white-poplar. *See* Tulip-Poplar
white walnut. *See* Butternut
whitewood. *See* Tulip-Poplar
whorled leaf arrangement, 3–4, 11, 328
wild banana. *See* Pawpaw
wild black cherry. *See* Black Cherry
wild-olive. *See* Devilwood
wild plum. *See* Chickasaw Plum
Willow Oak (*Quercus phellos*), 22,
  167–168
  bark, 168
  bark of large tree, 167–168
  fruit, 167
  leaves, 167
  staminate flowers, 167
  winter twig, 315
Willow or Poplar Family (*Salicaceae*),
  14, 15, 259–262
  Black Willow (*Salix nigra*), 28,
    261–262
  Eastern Cottonwood (*Populus deltoides*), 28, 259–260
  guide to, 28
  winged, 328
Winged Elm (*Ulmus alata*), 30, 303–304
  bark of large tree, 304
  bark of small tree, 304
  flowers, 303
  fruit, 303
  leaves and winged twig, 303
  winter twig, 322
Winged Sumac (*Rhus copallinum*), 17,
  63–64
  bark, 63–64
  flowers, 63
  fruit, 63
  leaves, 17, 63
  winter twig, 309
  winter twigs, 309–322
Witch-Hazel (*Hamamelis virginiana*)
  unique stringy yellow petals of flowers, 82

wavy margin and lopsided base of
  leaves, 82
Witherod (*Viburnum cassinoides*), 58
woolly buckthorn. *See* Gum Bumelia
wooly, 328

xylem, 5

Yaupon (*Ilex vomitoria*), 77–78
  bark, 78
  lanceolate Myrtle-Leaved Holly leaves
    for comparison, 78
  leaves and flowers, 77
  leaves and fruit, 17, 77
  mostly entire margin of Dahoon leaves
    for comparison, 78
yellow ash. *See* Black Locust
yellowbark oak. *See* Black Oak
Yellow Birch (*Betula alleghaniensis*)
  golden bark of a young tree, 84
Yellow Buckeye (*Aesculus flava*), 28,
  263–264
  bark, 264
  flowers (courtesy of John Seiler), 263
  fruit, 263
  leaf, 263
  seeds showing "buck's eye," 263
  winter twig, 320
yellow buckthorn. *See* Carolina
  Buckthorn
yellow-flower magnolia. *See*
  Cucumbertree
yellow locust. *See* Black Locust
yellow oak. *See* Black Oak; Chinkapin
  Oak
yellow-poplar. *See* Tulip-Poplar
yellow wood. *See* Sweetleaf
Yellowwood (*Cladrastis kentukea*)
  obovate leaflets, 126
  winter twig, 312

Zanthoxylum clava-herculis. See
  Hercules'-Club